Evolution of the Universe

Evolution of the Universe

I.D. NOVIKOV
SPACE RESEARCH INSTITUTE, MOSCOW

TRANSLATED BY M.M. BASKO

CAMBRIDGE UNIVERSITY PRESS

CAMBRIDGE
LONDON   NEW YORK   NEW ROCHELLE
MELBOURNE   SYDNEY

Published by the Press Syndicate of the University of Cambridge
The Pitt Building, Trumpington Street, Cambridge CB2 1RP
32 East 57th Street, New York, NY 10022, USA
296 Beaconsfield Parade, Middle Park, Melbourne 3206, Australia
Originally published in Russian as *Evolyutsiya Vselennoï* by Nauka,
Moscow, 1979 and © Glavnaya redaktsiya fiziko-matematicheskoi
literaturi, izdatelstva 'Nauka' 1979
First published in English by Cambridge University Press 1983 as *Evolution
of the Universe*
English edition © Cambridge University Press 1983

Printed in Great Britain at the University Press, Cambridge

Library of Congress catalogue card number: 82-9475

British Library Cataloguing in Publication Data

Novikov, I.D.
Evolution of the universe.
1. Cosmology
I. Title II. Evolyutsiya Vselennoï. *English*
523.1   QB981

ISBN 0 521 24129 4

Вселенная тоже была молодою
И бился в груди её пламень творенья.
Как женщина, власть потеряв над собою,
Она отдавалась на волю мгновенья.

И в огненной пляске Пространства и Времени,
Доверившись слепо неведомым силам,
Она разрешилась от тяжкого бремени,
Даруя начала Мирам и Светилам.

Дыханье горячее Тайны Великой
Потоками квантов к тебе прикоснётся,
И Космос огромный, чужой, многоликий
Сквозь Мрак Мирозданья тебе улыбнётся.

И тот, кто увидел улыбки той отблеск,
Кто вздрогнул на миг и застыл ослеплённый,
Тот будет всю жизнь, позабыв сон и отдых,
Искать её снова в просторах Вселенной!

Б. Комберг

The Universe once was also young
And Her heart was in the flame of creation
Like a woman having lost control of herself
She gave in to the violent burst of expansion

In a fiery dance of Space and Time
With blind obedience to the laws of the Unknown
She gave birth in labour and pain
To the host of worlds, and the Sun with the Earth – our home

When the heat from a breath of the Greatest of Mysteries
Will whiff in your face with the flows of quanta,
You will probably catch – through the darkness of skies –
The miraculous smile of the vast and impassionate stranger – the
Cosmos

And once you have noticed the gleam of that smile
And started, and stood for a moment all struck with amazement,
You will never forget and will spend all your life
In an anxious search for yet another glimpse of that vision.

B. Komberg
(As translated by M.M. Basko)

Contents

ix

Preface

There is nothing more grandiose than the global evolution of the entire Universe. No less striking is the power of reason apprehending this evolution. 'The most incomprehensible thing about the world is that it is comprehensible' wrote Einstein.

Human reason – like everything in nature – strives for extremes. And no wonder that many people all over the world wish to know more about the most extreme of all mysteries – the mystery of the entire Universe.

The aim of this book is to satisfy such a need. But, at the same time, I have set as my goal the task of showing how this mystery of mysteries is being unravelled by modern science, with its reliable and verified methods. In this way cosmology is being deprived of some mystic aura, and its conclusions appear as certain as the conclusions of other branches of natural science.

I am grateful to my English colleagues for issuing an English edition of this book and making it accessible to a large circle of readers. I am especially pleased with the fact that it is being published in England. The scientists of this country – beginning with the great Newton – made a major contribution to our present-day knowledge of the Universe.

The majority of facts presented in the book are firmly established and the results concerning the evolution of the Universe are not liable to revision. Therefore, although several years have elapsed since the publication of the Russian edition, the text of the book does not need much revision. In fact, all the latest achievements in cosmology up to the spring of 1981 have been included in this edition. During these years the general trend in contemporary cosmological research, described in the book, has become even clearer: investigations are becoming more and more concentrated around two big problems – the mystery of the beginning of the expansion of the

Universe and the riddle of the origin of galaxies and the large-scale structure of the Universe.

To make the presentation of material clearer than in the first Russian edition, parts of the original text have been rewritten. I am very grateful to Dr. M.M. Basko, the translator of the book, who is a well-known Soviet specialist on relativistic astrophysics and cosmology. His comments and advice contributed noticeably to the improvement of the English edition of the book. I wish to express my gratitude also to all my other colleagues whose comments helped me to improve this edition.

The translator and I are very much obliged to L.G. Straut who expertly managed the typing of the manuscripts, both in her native Russian and alien English.

Moscow, April 25, 1981   I.D. Novikov

Preface to the Russian edition

Cosmology is the study of the structure and evolution of the Universe. It is, in essence, a branch of natural science that is doomed to remain at the junction of different fields of knowledge. The latter means that cosmology incorporates ideas and methods of quite different character. Being a branch of astronomy, cosmology at the same time employs the methods and achievements of physics, mathematics and philosophy. The subject of cosmology is all the surrounding macrocosm, the entire 'Grand Universe'; its goal is to construct physical and mathematical models describing the most general properties of the structure and evolution of the Universe. Clearly, the basic conclusions of cosmology are of major significance for the *Weltanschauung* in general. It is important to emphasize that the theoretical models of modern cosmology are verified by the methods of observational astrophysics, i.e. according to the criteria of practice and experience.

In this book the achievements of modern cosmology are presented in such a form that they can be understood by the layman, and not just by the specialists to whom books on cosmology are usually addressed.

The reader is assumed to have no special knowledge beside the general information acquired from school astronomy, physics and mathematics courses. Nevertheless, the seriousness of the problems discussed demands some effort and thought on the part of the reader.

Before applying myself to the writing of this book, I, together with Ya.B. Zel'dovich, had written two monographs on modern cosmology for specialists (*Relativistic astrophysics*, Nauka, Moscow, 1967 and *Structure and Evolution of the Universe*, Nauka, Moscow, 1975). In these books the investigations by the authors and their colleagues have been summed up and the achievements of the whole of cosmological science have been surveyed in general.

Naturally, when discussing many issues in this present book, I have followed the above mentioned monographs and some other works with Ya.B. Zel'dovich and collaborators – sometimes just omitting complicated mathematical calculations and presenting the material in a somewhat simpler way. As a matter of fact, my goal was to recount the original works so that they became intelligible to a wide circle of readers. Our popular articles have been used in this book too.

The author is grateful to Yy.N. Efremov and O.Yu. Dinariev who read the manuscript and made a number of valuable comments, and to L.G. Straut who did all the typing.

Moscow, January, 1979 I.D. Novikov

Introduction

Nowadays the idea of the evolution of the Universe appears quite natural and even necessary, but this has not always been so. Like any great scientific idea, it took a lot of shaping and struggling through before it eventually triumphed in science. Today, the evolution of the Universe is a scientific fact, substantiated by many astrophysical observations and supported by the solid theoretical basis of all physics.

This book describes the evolutionary Universe and cosmology – that branch of natural science whose subject is the Universe as a whole.

Physical cosmology as a true branch of science may be considered as an achievement of the twentieth century. It was only in this century that Albert Einstein created the relativistic theory of gravity (the general theory of relativity) which serves as the theoretical basis for the science of the structure of the Universe. On the other hand, the achievements in observational astronomy at the beginning of this century – the verification of the nature of galaxies, and the discovery of Hubble's red shift law – together with the extraordinary advances in radio astronomy and in other methods of observation (such as those involving spacecraft), have in recent decades established a firm experimental basis for modern cosmology.

The beginning of modern cosmology is marked with the works of the brilliant Soviet scientist A.A. Friedmann, written in the years of 1922–4. On the basis of Einstein's theory he constructed mathematical models of the motion of matter in the whole of the Universe under the action of gravity. Friedmann proved that the matter in the Universe cannot be at rest: the Universe cannot be stationary – it must either expand or contract and, consequently, the density of matter in the Universe must either decrease or increase. In this way the concept of the global evolution of the Universe had been developed theoretically.

1

This idea was utterly new and very unusual. Various models of the structure of the Universe, succeeding one another, had prevailed in science for centuries. But all of them (or nearly all of them) had one feature in common – they were models of just the structure and not of the evolution and continuous change of the Universe; models of a sort of invariable 'clockwork Universe'. The reigning idea of a stationary (or steady-state) Universe seemed self-evident. The most complicated processes might occur in the Universe, but from where – from which state – and to where could the entire Universe evolve? The idea of the evolution of the whole of the Universe appeared absurd, preposterous, and it took much effort even for outstanding scientists to accept this idea. As an example, Albert Einstein himself may be mentioned. The creator of the theory of relativity was fully aware of how important his theory was for cosmology. Immediately after the equations of the general theory of relativity had been written, he tried to find out whether they admitted *static* solutions when applied to the whole of the Universe, i.e. solutions describing a state which does not change with time. It seemed quite evident to Einstein that it was a static, non-evolving model of the Universe which was to be constructed. But the equations of the general theory of relativity, when applied to the entire Universe, allowed no static solutions. The equations of the new physical theory proved to contain more than their author could anticipate. Note, that this is true not only for general relativity but for any kind of really significant theory. The idea of the static Universe seemed so attractive that Einstein did not believe his equations and began to change them.

Why then did the idea of the static Universe seem so attractive? The reason for this apparently stems from the fact that all the astronomical bodies and systems – be they the Solar System, stars, stellar clusters or galaxies – undergo no visible change, they remain always the same. Willingly or unwillingly, the observed invariability of astronomical phenomena over all scales accessible to mankind had been extrapolated to the whole of the Universe. It was quite clearly formulated by Aristotle in his treatise 'On the Heavens': 'Throughout all past time, according to the records handed down from generation to generation, we find no trace of change either in the whole of the outermost heaven or in any one of its proper parts.'

From today's point of view, that anti-evolutionary prejudice, that search for static solutions to cosmological equations appears to be wrong even in principle because the evolution of all those heavenly bodies and systems, which earlier seemed to shine constantly and move steadily along circular orbits, has now been firmly established. We can remind ourselves that the mere fact of the irreversibility of radioactive decay provides good evidence for evolution. If the Earth – a heavenly body – existed for an infinite time, all the radioactive elements would have long (infinitely long!) since decayed. But we know that radioactive elements are present in the terrestrial crust. Hence, some finite period of time has elapsed since the Earth, its crust and radioactive elements formed. Now even schoolchildren know that the amount of radioactive substance in rocks serves as a good indicator for the age of the Earth.

The evolution of the Sun and the stars may be considered as firmly established too. These celestial bodies radiate energy. The source of their energy is the nuclear reactions in stellar interiors. Any source of energy can be exhausted. The stores of nuclear energy are not unlimited either. Hence, the Sun and the stars formed at some finite point in the past and have their own history.

Today, we observe violent explosive and evolutionary processes in such gigantic systems as the galaxies too. The matter constituting the galaxies is being gradually reprocessed via nuclear reactions occurring in stars. Hydrogen is being converted into helium, and into heavier elements afterwards.

Thus, a static picture is unacceptable for any kind of astronomical object, if only long enough periods of time are considered. If we were just now to begin the construction of the model of the Universe, we would be compelled to develop an evolutionary model – a model in which the epoch was to be specified when the formation of stars, galaxies, etc. commenced in the Universe.

Let us return, however, to the beginning of this century. Einstein succeeded in the construction of a static model of the Universe. But the price for that was a hypothetical force of cosmic repulsion which he had to introduce into his theory in addition to the force of universal gravitational attraction. When later he became acquainted with the paper of A.A. Friedmann, who proved that the Universe must be non-stationary, he decided that this paper was incorrect. And only after the explanations of Friedmann had been conveyed to Einstein, did he recognize the accuracy of the conclusions obtained by the Soviet mathematician and acknowledge his own attempt to construct a non-evolutionary model as a faulty one. And at last, in 1929, the American astronomer E. Hubble, having analysed the results of many observations, established the fact of the expansion of the Universe.

In this way the global evolution of the Universe was proved. This discovery became one of the greatest achievements of human reasoning. But the discovery of the law of expansion of the Universe was, of course, just the very beginning of the exploration of its evolution. This was, so to say, the verification of just the mechanics of the Universe. There still remained to be investigated the specific *physical* processes occurring in the expanding Universe – the processes that occurred in the remote past, when the state of matter strongly differed from the present one, as well as those that occurred in epochs closer to modern times, when the heavenly bodies and their systems began to form, and, finally, those that occur today and are to occur in the future.

It was just in the last ten to twenty years of the development of cosmology that the mechanical picture of forces and motions in the Universe has been given a new concrete physical meaning.

The subsequent chapters of this book expound the contemporary scientific view of the evolution of the Universe. And the picture of this evolution looks really grandiose: the expansion from a superdense and superhot state with violent reactions between elementary particles to its contemporary state with matter fragmented into gigantic systems of heavenly bodies, where stars and planets have formed and life has developed.

It is to be emphasized, however, that this book is about the Universe itself and about the development of new ideas concerning the Universe rather than specifically about the people working and having worked in this branch of science. To try to discuss both these aspects simultaneously does not seem right to me because they are different tasks which require different styles. So, in the subsequent discourse I do not follow the chronological order and only occasionally mention names and dates.

Finally, note that the term 'Universe', as generally accepted in cosmology literature, is used below to denote all the surrounding macrocosm. In different kinds of scientific literature – such as literature on philosophy, physics and astronomy – sometimes different meanings are attributed to this term. Throughout the text concrete physical properties and physical and mathematical models of the surrounding macrocosm are discussed.

The reader who would like to have a more detailed and deeper understanding of the problems of modern cosmology is referred to the following specialist books:

D. Sciama, *Modern Cosmology*, Cambridge University Press, 1975;
P.J.E. Peebles, *Physical Cosmology*, Princeton University Press, 1971;
S. Weinberg, *Gravitation and Cosmology*, New York, Wiley, 1972;
Ya.B. Zel'dovich & I.D. Novikov, *Structure and Evolution of the Universe*, Moscow, Nauka, 1975 (English translation in preparation).

My point of view on the problems of cosmology is presented in the last book from the above list. Since our book appeared, many specific issues underwent further development, but the overall view of general problems has remained essentially unaltered, and I followed the above mentioned book when discussing many issues (see the Preface).

1

The expanding Universe

1   The large-scale homogeneity and isotropy of the Universe

Any attempts to construct a model of the world that surrounds us begin, of course, with the efforts to comprehend the observations.

What is the Universe that we see?

In this introductory section we discuss only the most general conclusions from the observations which are important for understanding the following text. A more detailed review of the observational data is given in §§ 8 and 9 of this chapter.

Until recently astronomers could directly observe only shining bodies, i.e. stars, luminous gas and stellar systems.

On a comparatively small scale the distribution of stars in space is quite non-uniform. This became clear when it was realized that the Milky Way was a gigantic cluster of stars – the Galaxy. As the power of optical telescopes increased and the astrophysical methods became more sophisticated, it turned out that the galaxies are quite numerous and rather inhomogeneously scattered in space, and that the Universe in general appears as an aggregation of clusters of galaxies. The clusters of galaxies differ widely in size and in the number of individual members. Big clusters contain thousands of galaxies and extend to megaparsecs (Mpc) in size. (Astronomers measure distances in parsecs (pc): 1 pc = 3.26 light-years = 3.1×10^{18} cm. In cosmology 1 megaparsec = 10^6 pc is used as a unit length.) On average, they are some 30 Mpc apart, i.e. their separations are typically about 10 times as large as their dimensions. The latter implies that the average density within these structural units is 100–1000 times greater than the density that one would obtain, having spread all the matter uniformly in space. There are even bigger condensations – the superclusters. Thus, below a scale of 30–100 Mpc the Universe 'splits' into individual structural units and is

7

inhomogeneous. But if one were to cut out a cube with an edge 10 times as long, it would always contain about the same number (~ 1000) of galaxy clusters, irrespective of its location in the Universe, i.e. the Universe is approximately homogeneous on larger scales. While galaxy clusters were studied with only optical instruments, we could not penetrate deeply enough into space because the telescopes did not permit investigation of even the brightest objects at distances in excess of a few billion parsecs. Such a volume contains about a million clusters of galaxies. The precision of optical methods in determining the spatial distribution of galaxies is not too high, and the large-scale homogeneity of the Universe could be ascertained to an accuracy of 10–20% only. Over the last two decades new methods enabling us to explore the large-scale homogeneity and isotropy (meaning the independence of some properties on direction in space) of the Universe have been developed. First of all, they arose from the discovery of the cosmic microwave background – the relict radiation, which comes to us from huge distances and which we discuss in greater detail below. Here we note only that even the most precise latest measurements could not detect any difference in its intensity from different directions in the sky in excess of one part in 10^3–$10^{4\dagger}$. The latter fact provides compelling evidence that the properties of the Universe are the same in all directions, i.e. the Universe is isotropic to a high degree of accuracy. And these same data imply also – as we shall see below – that the Universe is homogeneous to a high degree of accuracy. Deviations of the matter density from its average value on a scale of 1000 Mpc do not exceed 3%, and are even less on larger scales.

Thus, one of the basic observational features of the Universe is its inhomogeneous structured appearance on small scales, and quite perfect homogeneity on large scales.

On scales of hundreds of megaparsecs the matter in the Universe may be envisaged as a homogeneous continuous medium whose 'atoms' are galaxies or clusters of galaxies.

In the last century attempts have been made to construct the so-called hierarchic models of the Universe. These models pictured

---

† We do not discuss here the small intensity difference of the relict radiation, coming from two opposite directions in the sky, which is caused by the motion of the Sun with a speed ~ 390 km s^{-1} relative to the system of all other galaxies.

the Universe as an infinite sequence of hierarchic structures, each of a successively higher order than the previous one: the stars aggregate into galaxies, the galaxies group into clusters, the clusters make superclusters and so on, up to infinity. Such a picture, however, does not conform to the observations.

When studying the large-scale structure of the Universe, one must proceed from its basic properties of homogeneity and isotropy.

2 Theory predicts a non-stationary Universe

Let us see what conclusions can be drawn from the knowledge that matter in the Universe is uniformly distributed.

The most important force acting on celestial bodies is the force of gravitational attraction. The basic law governing the action of this force was established by Isaac Newton in the seventeenth century.

The Newtonian theory of gravity and Newtonian mechanics became the greatest achievements of the natural sciences. They provide a very accurate description of a wide range of phenomena, including the motion of natural and artificial bodies in the Solar System, as well as the motions in other cosmic systems – binary stars, stellar associations, galaxies.

On the basis of Newtonian gravity theory a number of impressive predictions were made – such as the prediction of a previously unknown planet Neptune, the prediction of a companion star to Sirius, and many others that were brilliantly confirmed later. At present, Newton's law serves as a foundation for all calculations of motion, structure and evolution of celestial bodies and for determination of their masses. But in some cases, when gravitational fields become strong enough and the velocities of moving particles approach the speed of light, gravitation cannot be adequately described by Newton's law. In such cases one must resort to the relativistic theory of gravity created by Albert Einstein in 1916.

The necessity to overgo the framework of Newtonian gravity in the cosmological problem was realized long ago, many years before Einstein created a new theory. We shall return to a more detailed discussion of this later, but intuitively it is clear that to calculate the gravitational field of an infinitely massive Universe one might need something more general than Newton's law.

It turns out, however, that both Newton's and Einstein's theories of gravity have an important feature in common which enables one, using Newton's theory alone, to elucidate the most important property of a homogeneous Universe – its non-stationary character. We shall turn to Einstein's theory later on and discuss some of its conclusions that are necessary for the construction of complete self-consistent cosmological models and for the analysis of physical processes in the evolving Universe. Such an excursion, as has been already emphasized, is absolutely unavoidable and will be justified below. But if the basic property of the Universe – its dynamic behaviour – can be understood and described in the framework of Newtonian gravity, then invoking Einstein's theory would lead to an unnecessary confusion of the problem.

Now we come back to the important common property of Einstein's and Newton's theories which reads as follows: a hollow spherically-symmetric shell of matter does not create any gravitational field in its interior. We shall demonstrate this in the Newtonian case.

Consider a thin spherical shell of matter as shown in Fig. 1. We are going to compare gravity forces that pull a particle of mass m (located at an arbitrary point inside the shell) in two opposite directions a and b. The direction of the line ab, passing through m, is supposed to be arbitrary too. The forces of gravitational attraction are created by the matter within the two surface elements cut out from the shell by two narrow cones with equal vertex angles. The areas of the surface elements, cut by these cones, are proportional to the squares of the cone heights. Namely, the ratio of the area S_a of an element a to the area S_b of an element b is equal to the ratio of the squares of the

Fig. 1. The particle m is attracted to surface elements a and b with gravity forces that are equal in magnitude and opposite in direction.

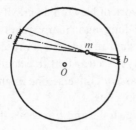

distances r_a and r_b from m to the shell surface along the line ab:

$$S_a/S_b = r_a^2/r_b^2 \qquad (1)$$

Since the mass is supposed to be evenly distributed over the shell surface, we arrive at the same ratio for the masses of the surface elements:

$$M_a/M_b = r_a^2/r_b^2. \qquad (2)$$

Now we can calculate the ratio of the forces with which surface elements attract the particle. According to Newton's law the expressions for these forces are

$$F_a = GM_a\, m/r_a^2, \quad GM_b\, m/r_b^2. \qquad (3)$$

Their ratio is given by

$$F_a/F_b = M_a\, r_b^2/M_b\, r_a^2. \qquad (4)$$

Substituting for M_a/M_b in eq. (4) its value from eq. (2), we finally get

$$F_a/F_b = 1, \quad F_a = F_b. \qquad (5)$$

Hence, the two forces are equal in magnitude and act in opposite directions – thus cancelling each other out. The argument can be repeated for any other direction. As a result, all forces pulling m in opposite directions cancel one another out and the resultant force is exactly zero. The location of particle m was arbitrary. Hence, there truly are no gravitational forces inside any spherical shell.

Now let us turn to the effect of gravitational forces in the Universe. In the previous section the distribution of matter in the Universe was shown to be homogeneous on large scales. Throughout this chapter we discuss large scales only, assuming the matter to be uniformly distributed in space.

Let us single out of this uniform background an imaginary sphere of an arbitrary radius with the centre at an arbitrary point, as depicted in Fig. 2. Consider first the gravitational forces that the matter inside the sphere exerts on the bodies at its surface, ignoring for a moment the effect of matter outside the selected sphere. Let the radius of the sphere be not too large, so that the gravitational field generated by its interior is relatively weak and Newtonian theory of gravity applies. Then, the galaxies at the surface of the sphere are attracted to its centre by forces which are directly proportional to the mass M of the sphere, and inversely proportional to the square of its radius R.

Next, consider the gravitation effect of all the remaining material in the Universe, which lies outside the sphere under consideration. All this matter can be thought of as a sequence of concentric spherical shells with increasing radii, surrounding our selected sphere. But, as we have already shown, spherically symmetric layers of material create no gravitational forces in their interiors. As a result, all the spherical shells (i.e. all the remaining material of the Universe) add nothing to the net force attracting some galaxy A at the surface of the sphere toward its centre O.

So, we can calculate the acceleration of one galaxy A with respect to another galaxy O. We have associated O with the centre of the sphere, while A is at a distance R from it. The acceleration to be found is due to the gravitational attraction of matter inside the sphere of radius R only. According to Newton's law it is given by

$$a = -GM/R^2. \tag{6}$$

The negative sign in eq. (6) reflects the fact that the acceleration corresponds to attraction rather than to repulsion.

Thus, any two galaxies in a homogeneous Universe separated by a distance R experience a relative (negative) acceleration a as given by eq. (6). This means that the Universe cannot be stationary. Indeed, even if one assumes that at some instant all the galaxies are at rest and the matter density in the Universe does not change, the very next moment the galaxies would acquire some speed due to mutual

Fig. 2. The gravity force, attracting some galaxy A at the surface of a sphere of an arbitrary radius R toward its centre O, is determined by the total mass inside the sphere only, and is not affected by the material outside the sphere.

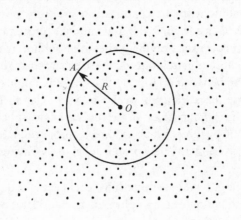

gravitational attraction resulting in the relative acceleration represented by eq. (6). In other words, the galaxies could remain motionless with respect to each other only for a brief instant. In the general case they must move – either approach each other or recede from each other; the radius of the sphere R (see Fig. 2) must change with time, and so must the density of matter in the Universe.

The Universe must be non-stationary because gravity acts in it – that is the basic conclusion from the theory. This conclusion, as has been already mentioned in the Introduction, was first reached by A.A. Friedmann in the framework of the relativistic gravitational theory in 1922–4. Some years later, in the mid-thirties, E.A. Milne and W.H. McCrea pointed out that the non-stationary behaviour of the Universe can be also derived from Newtonian theory as outlined above.

How specifically would we expect the galaxies to move, and the density to change?

This depends not only on the gravitational forces governing the motion. The forces determine the acceleration or, more precisely, the deceleration (the negative sign in eq. (6)) of matter; in other words, they determine the rate at which velocities change with time. Apparently, one has to know the values of the velocities at some moment to be able – with the acceleration known – to calculate how these velocities change with time. If one assumes the galaxies to be at rest at some instant, they begin to draw together at later moments and the Universe as a whole starts to contract. If one assigns an initial velocity distribution corresponding to all galaxies moving away from one another, one arrives at the expanding Universe whose expansion is decelerated by gravity.

The magnitude of the velocity at a given time cannot be deduced from the gravitational theory, it must be established by the observations.

3   **The discovery of the expansion of the Universe**

Remote stellar systems – galaxies and clusters of galaxies – are the largest structural units of the Universe known to astronomers. They can be seen from enormous distances, and it was the study of their motions that served as the basis for the exploration of the kinematics of the Universe. The velocities of distant objects moving toward or away from the observer can be measured with the aid of the Doppler effect. Recall that due to this effect all the wavelengths of the light coming from a source approaching the observer are shortened, or shifted toward the violet part of the spectrum; the light coming from a receding source, on the other hand, is reddened, or detected at longer wavelengths. The magnitude of the wavelength shift is usually denoted by the letter z and is defined as:

$$z = (\lambda_{obs} - \lambda_{em})/\lambda_{em} = v/c. \tag{7}$$

Eq. (7) is valid only for the velocities v much less than the speed of light c, when the laws of Newtonian mechanics are applicable. If the velocity v of the light source approaches the speed of light, the formula expressing z through v becomes more complicated and is not considered here, since for the time being we are going to discuss small velocities as compared to c.

By measuring the displacements of spectral lines in the spectra of celestial bodies, astronomers can determine whether they move toward us or away from us; in other words, astronomers measure the projection of their velocity along the line of sight – the radius connecting the observer with the object in question. This is why the velocities determined from the spectral measurements are called the radial velocities.

Pioneering work in measuring the radial velocities of galaxies had been undertaken at the beginning of this century by the American astrophysicist V.M. Slipher. At that time the distances of galaxies were not known, and it was a matter of hot controversy as to whether they belonged to our stellar system – the Milky Way – or not. Slipher discovered that the majority of galaxies (36 out of 41 studied) receded from us with velocities that range up to almost 2000 km s^{-1}. Only a few were approaching us. (As it turned out later these were the closest.) In the course of subsequent studies it has been established that the Sun orbits the centre of our Galaxy with a speed of 250 km s^{-1}, and that the 'approach velocities' of a small number of closest galaxies stem mostly from the fact that the Sun at present is moving toward these objects.

Thus, according to Slipher the galaxies were receding from us. The lines in their spectra were shifted toward the red. This phenomenon was called the 'red shift'.

In the 1920s the distances of galaxies were measured. This was accomplished with the aid of pulsating stars, whose brightness varies in a regular manner, – the Cepheid variables.

These variable stars have a remarkable property: the amount of light emitted by a Cepheid variable per unit time – its luminosity – is closely correlated with the period of its pulsations. Knowing the period, one can calculate the luminosity. The latter, in its turn, enables one to determine the distance. Thus, having measured the pulsation period of a Cepheid variable from its observed light variations, one can find its luminosity. One can then measure its apparent brightness. For a given luminosity the apparent brightness is inversely proportional to the square of the distance to the star. Comparing the apparent brightness with its luminosity, one can infer the distance to a Cepheid variable.

Cepheid stars were also discovered in other galaxies. The distances of these variables and, as a consequence, the distances of galaxies in which they reside turned out to be much greater than the size of the Milky Way – our own galaxy. In this way the galaxies had been finally proven to be remote stellar systems similar to the Milky Way.

Beside Cepheids, other methods of determining the distances of galaxies have also been used from the very beginning. One such method employs the brightest stars of galaxies as indicators of their distances. The brightest stars of all galaxies, including our own, seem to have approximately the same luminosity, thus providing us with a 'standard candle' which can be used to calculate the distances. The brightest stars have greater luminosities than Cepheid variables and can be seen at greater distances. As a result, they extend somewhat our means of measuring distances in the Universe.

The distances of a number of galaxies were determined by the American astronomer E.P. Hubble. Comparing the distances of galaxies with their recession velocities (velocities had been determined earlier by Slipher and other astronomers and were only being corrected for the motion of the Sun inside the Galaxy), in 1929 he found a remarkable correlation: the farther away the galaxy, the greater is its recession velocity. The recession velocities of galaxies turned out to be related to their distances by a simple law:

$$v = Hr. \tag{8}$$

The proportionality coefficient H is now called the Hubble constant.

Fig. 3 shows the plot of the recession velocities of the galaxies versus their distances from which Hubble deduced his law. According to this plot the Hubble constant is $H \approx 500$ km s^{-1} Mpc^{-1}. About 50 years have passed since Hubble discovered his law. The power of astronomical instruments and methods has increased tremendously, but all subsequent investigations only confirmed the Hubble law, eq. (8), – the recession velocities of the galaxies are proportional to their distances. Hubble, however, had greatly overestimated the value of the proportionality coefficient H. The current value of H is almost ten times less than that obtained by Hubble. We shall discuss the origin of this error in § 5, but now we return to the implications of Hubble's discovery.

This discovery has revealed that the galaxies move away from us in all directions with velocities that are directly proportional to their distances.

One's first impression is one of involuntary amazement: Why do all other galaxies run away just from us – from our galaxy? Are we really at the centre of the Universe?

Such a conclusion would certainly be wrong. The point is that the galaxies recede not only from our galaxy, but from each other too. If

Fig. 3. Radial velocity–distance relation for galaxies according to E.P. Hubble; the dashed line represents the data of 1929, the solid line the data of 1936. (It should be noted here that the scatter of individual points about the straight lines of the plot is not only due to inevitable observational errors, but is also a manifestation of the random motions of the galaxies superimposed on the smooth expansion velocity field.)

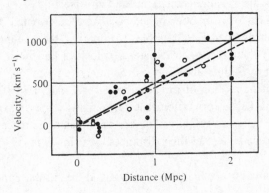

we happened to live in some other galaxy, we would see exactly the same pattern of receding galaxies as we do now. To clarify this, turn to Fig. 4. Assume first that we, the observers, are in galaxy A and consider it to be at rest (Fig. 4*a*). To simplify the argument, let us have a look at the galaxies along the horizontal line only. The galaxies B, $C, \ldots$ recede from us to the right with increasing velocities. The galaxies $D, E \ldots$ recede to the left. Now, let us move to galaxy B and consider it to be at rest (Fig. 4*b*). To find new velocities of all the galaxies relative to B, the velocity of B is to be subtracted from all velocities in Fig. 4*a*.

Now A recedes from B to the left with the same velocity as earlier B receded from A to the right in Fig. 4*a*. The recession velocity of D doubles, and so on. Galaxy C recedes from B with the half of its earlier velocity relative to A, but it is twice as close. As a whole, the recession pattern seen from B is just the same: the velocities of recession are proportional to the distances, with the same proportionality coefficient as found from A. For the sake of simplicity we considered the galaxies along one straight line only, but one can easily verify that the conclusion drawn remains valid in the most general case too: any observer in any galaxy would see the same picture, as if the other galaxies were fleeing just from him.

Indeed, having moved to galaxy B, one has to perform a vector subtraction of its velocity from the velocities of all the remaining galaxies in Fig. 4*a* (as is shown for galaxy F), in order to obtain their

Fig. 4. (*a*) The recession pattern of galaxies as seen from A. (*b*) The recession pattern of galaxies as seen from B (According to V. A. Bronshtan).

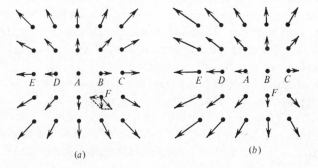

(*a*)   (*b*)

motion pattern with respect to *B*. As a result, one arrives at the picture given in Fig. 4*b*.

One can, perhaps, invoke an even simpler argument to verify that the expansion pattern described by Hubble's law appears the same for any observer at any point of space. Take a homogeneous sphere and, say, double its size in such a way that it remains homogeneous. Quite clearly, the separation of any two points inside the sphere will also double, regardless of where we have chosen them. Thus, an observer located at an arbitrary point inside an expanding homogeneous sphere will see the same picture everywhere with all the other points receding from him. If one now takes a sphere of an infinite size, one obtains exactly the picture described above.

So, the fundamental observational fact is that the galaxies recede from one another – the Universe is expanding. In this way, the theoretical prediction of a non-stationary Universe made by Friedmann has been remarkably confirmed.

Some people ask the following question. Suppose that on average the clusters of galaxies uniformly fill the whole of the Universe (we shall see later that there are forceful arguments in favour of such a hypothesis), 'into what', then, does the Universe expand? Such a question, however, is incorrectly stated. The Universe itself is all that exists. Nothing can be found outside or beyond the Universe. Nothing means not only no galaxies or some other form of matter; nothing means nothing at all – no space, no time. There is no such a void into which anything could expand. But there is no need for anything beyond the Universe for it to expand into. One can try to illustrate this with the following example. Consider an infinite plane, over which 'dots' (i.e. the galaxies) are uniformly distributed. Let us now stretch this plane uniformly in all directions, so that all the distances between the dots increase. One could ask: to where was the plane stretched? But its extensions were already infinite from the very beginning. Such are the properties of infinity. On increasing infinity by a factor of two, we still get infinity!

One more important question naturally arises: why *does* the Universe expand? What was it that imparted velocities to the galaxies? Once again neither Newtonian nor Einstein's gravity theory can answer this question. The galaxies at present are moving by inertia, while their motion is decelerated by gravity; and it is only this deceleration – as given by eq. (6) – that the gravitational theories predict. We shall come back to this issue again in Chapter 5.

To conclude this section, we wish to make the following point. Sometimes one hears the statement that the expansion of the Universe causes everything to expand; the galaxies not only recede from one another, but they themselves expand; the stars are expanding, the Earth – just everything. This is certainly not true. The recession of galaxies has no effect whatsoever on individual bodies. We saw before that a homogeneous infinite medium generated no gravitational field inside a spherical cavity, i.e. did not influence individual bodies. Just as in an expanding gas cloud individual molecules do not expand, the gravitationally bound bodies – galaxies, stars, the Earth etc. – do not expand in the course of the expansion of the Universe. One can put forward an even stronger statement: if absolutely all bodies, including atoms, participated in the cosmological expansion in proportion to their sizes, such an expansion would be completely undetectable. There would simply be no unchanging rulers to measure the magnitude of expansion with. Of course, they can sometimes expand or contract, but due to inherent properties of their own, as a result of processes occurring in their interiors.

4 Superluminal velocities of galaxies?

According to the Hubble law (see eq. (8)) the recession velocities of galaxies are directly proportional to their distances and, as a result, there must exist a distance r_1 at which the recession velocity becomes equal to the speed of light; at still greater distances the velocities of galaxies will exceed the speed of light. Can this be true? No, it certainly cannot.

As a primary step in verifying the Hubble law, the red shifts of spectral lines in galactic spectra are measured. And only then are their recession velocities v calculated from eq. (7). But we have already mentioned that eq. (7) is valid only for velocities that are small compared to the speed of light. If the source of light has a velocity close to the speed of light (but, as before, there are still no strong gravitational fields), then a more complicated formula of special relativity theory,

$$v = c(z^2+2z)/(z^2+2z+2),$$ (9)

must be adopted to calculate v from z. Whatever the value of z (according to the definition $z \equiv (\lambda_{obs} - \lambda_{em})/\lambda_{em}$ it varies within the interval $-1 < z < +\infty$); the velocity v never exceeds c. If some object – say, in our galaxy – moved away from us with a speed close to c, then, having measured the red shift of its spectral lines z, one could calculate from eq. (9) its velocity. And the latter, surely, would never exceed c.

But for very distant galaxies with large red shifts even formula (9) becomes inapplicable. The fact is that eq. (9) does not account for the effect of gravity on the frequency change of the light waves. We shall see later that in cosmological models with strong and variable gravitation fields the usual simple meanings of the concepts of distance and velocity alter radically. It turns out that the very question – what is the velocity of a very distant galaxy relative to us? – is incorrectly stated. But we shall come to this later, and now turn to the Hubble constant.

5   Hubble constant

Hubble established his law by determining the distances of galaxies and by calculating their velocities from the measured values of the red shift z. In the first paper by Hubble of 1929 the recession velocities ranged up to 1200 km s^{-1}, which corresponds to a red shift of $z \approx 0.004$. We know now that the galaxies studied by Hubble are in our immediate vicinity.

Astronomers have been trying, of course, to verify Hubble's law at greater distances as well. For this one needs much more powerful indicators of distance than the Cepheid variables or the brightest stars discussed in §3.

In 1936 Hubble proposed to use entire galaxies as such indicators. His reasoning was as follows. Objects chosen as distance indicators must all have about the same luminosity. Then their apparent brightness will be strictly correlated to their distance. Single galaxies cannot be used in this way since their intrinsic luminosities strongly differ from one another. Our galaxy, for instance, radiates about the same amount of energy as ten billion suns. There are galaxies whose emissions are hundreds of times less, and there are galaxies, too, tens of times as bright. Suppose that there exists a fixed upper boundary to the total luminosity of individual galaxies. Then, in any rich cluster of galaxies, containing thousands of members, the brightest galaxy will most probably have a luminosity very close to this upper bound. In other words, the brightest galaxies of rich clusters should have about the same luminosity from cluster to cluster and may be chosen as a kind of standard analogous to Cepheid variables. The apparent brightness of these galaxies can be used as an indicator of distance. The farther away the cluster is, the weaker the brightness. And even if the exact value of the luminosity of the brightest galaxy is not known and one cannot calculate the distance itself, one can still check the form of the relation given by the Hubble law, (8) – the direct proportionality between the distance and the recession velocity – although the exact value of the coefficient H will remain uncertain.

Beginning from Hubble's paper, in order to test the validity of Hubble's law, astronomers usually plot the red shifts of clusters versus the brightnesses of their brightest members instead of plotting the red shift versus distance.

Hubble himself used the fifth brightest galaxy in every cluster. Note, however, that in astronomy the brightness of celestial objects is measured in stellar magnitudes. These units are purely historical in origin. The 1st magnitude had been assigned to the average brightness of the twenty brightest stars in the sky; the stars of the 2nd magnitude were then defined as being 2.512 times fainter; the 3rd magnitude 2.512 times weaker than those of the 2nd, and so on. The faintest stars that can be still discerned by the naked eye are of the 6th magnitude.

Of course, through large telescopes the spectra of stars and galaxies much fainter than those of 6th magnitude can be obtained. The largest telescope, the 6-metre of the Special Astrophysical Observatory at the Northern Caucasus, makes it possible to register objects of 24th magnitude.

Thus, in cosmology the relation of galactic stellar magnitudes m to their red shifts z is analysed. Such a relation, as taken from the paper by A. Sandage (1972), is shown in Fig. 5. For comparison, a black rectangle is shown in the lower left corner of the diagram. It corresponds to the data range covered by Hubble in 1929, when he discovered his law. We see that the progress in observational methods is very impressive. But until quite recently it seemed that advances would be even greater. These hopes were associated with powerful sources of radio waves that had been discovered with the aid of radio telescopes at distances exceeding those of galaxies. These radio sources, however, were found to undergo strong evolutionary changes and could not be used as 'standard candles'.

At the beginning of the 1960s quasars were discovered. Their optical luminosities are 10–30 times greater than those of galaxies, but they also turned out to be of no use in the problem of the cosmological expansion: in contrast to the brightest galaxies in clusters which all have about the same luminosity, luminosities of quasars are scattered over a wide range and, moreover, seem to vary strongly with time.

The plot in Fig. 5 shows that the Hubble law is valid up to the farthest distances measured.

The largest red shift measured up to now for galaxies is close to 1 and for quasars $z = 3.5$.

To transform the red shift versus apparent stellar magnitude relation into Hubble's law, (8), one has to determine the luminosity of the brightest galaxy in at least one cluster of galaxies. This is

Fig. 5. The relation between apparent stellar magnitude and red shift multiplied by the speed of light in km s^{-1} for the brightest galaxies in clusters (according to A. Sandage, 1972).

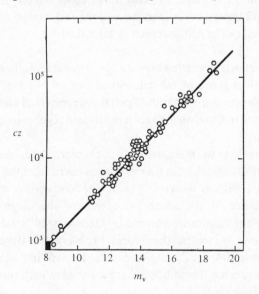

equivalent to the determination of the distance of a given cluster. But to find the distance of even one of the closest clusters – that in the Virgo constellation for example – presents a formidable task. We shall outline briefly how this has been done, omitting technical details. The method consists of a number of successive steps giving the distances of more and more remote objects, until one reaches the Virgo cluster.

By measuring trigonometric parallaxes, or angular displacements of stars over the celestial sphere due to the annual motion of the Earth around the Sun, one can determine only the distances of the nearest stars, no farther than 30 parsecs. This is not far enough to get the distances of even the closest star clusters. The determination of the distances to stellar clusters is the first step in the 'staircase of distances' leading eventually to the clusters of galaxies and to the establishment of the length-scale of the Universe.

From among the numerous star clusters one usually starts with the Hyades. This cluster (and others as well) moves as a whole with respect to the Solar System. By measuring the angular displacements of individual stars in the cluster over the celestial sphere, resulting from their common motion in space, one can obtain the value of angle α between the line of sight passing through any cluster star and the direction of cluster motion relative to the Sun. (This angle is equal to the angular separation of two points on the celestial sphere: one the position of a chosen star, the other the point where all apparent routes of individual stars converge.) Then one makes use of the radial velocities v_{rad} of the same stars, which enable one to calculate the tangential velocity v_t of the cluster as projected onto the celestial sphere: $v_t = v_{rad} \tan \alpha$. Now comparing this velocity with the rate of angular displacement β, one arrives at the value of the distance $R = v_t/\beta$. In this way, the distance of the Hyades cluster was shown to be 41 pc (1.3×10^{20} cm). With the distance known, one measures the apparent magnitudes and calculates the absolute luminosities of those cluster stars that are at present at the longest stage in their evolution – the stage of hydrogen burning (such stars are called 'main-sequence' stars – the term referring to their position on the magnitude–colour diagram). The luminosities of the main-sequence stars are closely correlated to their colours (the temperatures of their surfaces), and having established the distance of the Hyades cluster one can consider the luminosities to be known.

As the next step, one determines the distances of other, more remote star clusters, by comparing the apparent stellar magnitudes of the main-sequence stars in these clusters with those of Hyades' stars of the same colour. In this way one obtains the distances of all the star clusters in our galaxy. Some of these clusters contain periodic variable stars of a special class – the Cepheid variables – whose luminosities can then be easily calculated. The Cepheid variables – already discussed to some extent in § 3 – obey a remarkable relation connecting their luminosities M with the periods P of their light variations. Having established the luminosities M of those that are in star clusters, one finds the 'zero point' of the dependence $P–M$, i.e. a specific value of M corresponding to a specific value of P. The brightest Cepheids can be seen from large distances, up to 4 million parsecs away. They have been discovered in our neighbour, the spiral galaxy M31 (the famous Andromeda Nebula; the galaxies are labelled here according to one of the first catalogues – the Messier catalogue) and in some other close galaxies. Having measured the period of a Cepheid in M31, one finds its luminosity (from the $P–M$ relation) and comparing M with m, one evaluates the distance of the galaxy. Unfortunately, the Cepheids cannot be seen in the more distant galaxies of the Virgo cluster, which is significantly farther away than the Andromeda galaxy. One has to introduce an extra step. The galaxy M31 contains a lot of globular clusters (spherical star clusters). With the distance of M31 already known, one can calculate the luminosity M of the globular clusters. The globular clusters can be seen in galaxy M87, belonging to the Virgo cluster of galaxies. By measuring the apparent magnitudes of globular clusters in M87 and by comparing them with the value of M derived for these objects, one obtains the distance of the Virgo cluster (about 17×10^6 pc). Finally, using this distance and the apparent magnitude of the brightest galaxy in the cluster, one evaluates the luminosity M of this galaxy. It exceeds the luminosity of our galaxy (which is about 10^{44} erg s^{-1}) by approximately one order of magnitude.

Beside the above method, invoking Cepheid variables and globular clusters, other bright objects with more or less reliably determined luminosities are used as indicators of distances to galaxies. The list of these objects includes, for example, the brightest visible stars, novae and supernovae. In addition, the linear sizes of the regions of ionized hydrogen (HII regions) can be also used. They do not show much variation and may thus serve as indicators of distances too.

Fig. 6 shows the 'staircase' of different methods of measuring cosmic distances.

Obviously, errors can be introduced at any step of the long 'staircase' described above. For this reason, even now the coefficient H may be regarded as established with an accuracy probably not higher than a few tens of per cent. Recall that Hubble's estimate for H was about 500 km s^{-1} Mpc^{-1}. Contemporary estimates by Sandage, Tamman, de Vaucouleurs, van den Bergh and others give values ranging from 50 km s^{-1} Mpc^{-1} to 100 km s^{-1} Mpc^{-1}. What is the reason that Hubble made such a large error in estimating H?

The main sources of error were discovered only after 1950, when the 5-metre telescope on Mount Palomar, the largest telescope at the time, was set into operation. In 1952 the American astrophysicist W. Baade found that the Cepheid variables of the type used by Hubble were about four times brighter than had previously been thought. It meant that the distances of closest galaxies determined using Cepheids were in fact underestimated by at least a factor of 2. After some additional refinements, the distance of the Andromeda galaxy

Fig. 6. A 'staircase' of various methods for determining cosmic distances. The span of every line represents the range of distances covered by a particular method (according to S. Weinberg).

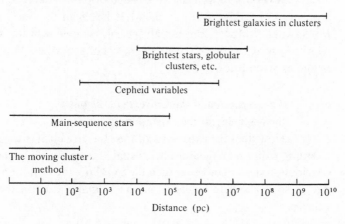

M31 turned out to be some 700 000 pc away, which was approximately three times farther than the first estimate by Hubble. As a result, the distances to all the other remote galaxies had to be multiplied by the same factor. This change in the distance scale caused the Hubble constant H to drop to a value of some 200 km s^{-1} Mpc^{-1}.

Before the above described revision of the distance scale all nearby galaxies seemed to be considerably smaller than our own. This appeared to be rather strange. After the revision it became quite clear that there are many galaxies of about the same size and even bigger than the Milky Way. Such a conclusion was an assurance that the revision of the distance scale was correct.

During the late 1950s it became evident that the distance scale for remote galaxies (in which no Cepheid variables could be seen), as determined by Hubble, was also in error. The reason was two-fold. First, some error had been introduced through the standard faint stars used by Hubble. Second, Hubble erroneously took the bright regions of ionized hydrogen (HII) in distant galaxies to be the brightest stars, which he used as 'standard candles'. The brightest stars turned out to be about five times less luminous than the HII regions. As a result, the distances of remote galaxies needed to be increased by an additional factor of 2.2.

In the end, the Hubble constant H turned out to be about 75 km s^{-1} Mpc^{-1}.

All of this revision had been accomplished by the early 1960s. The subsequent work in this area concentrated on the efforts to reduce the uncertainty in the above value of H. A laborious investigation of the problem by Sandage and Tamman has led them to conclude that $H \approx 55$ km s^{-1} Mpc^{-1}. But not all specialists agree with this value.

In this book, we use the value $H = 75$ km s^{-1} Mpc^{-1} for all the estimates given below.

6 The expansion of the Universe in the past; the beginning of the expansion

How does the expansion of the Universe change with time?

Again, as in §2, imagine a sphere isolated from the uniform matter distribution in the Universe, then try to follow the evolution of the size of such a sphere, while its surface remains fixed in the same galaxies.

The expansion is governed by the law of gravity. The acceleration (negative in sign – actually the deceleration) experienced by galaxies at the surface of a selected sphere is given by eq. (6)

$$a = -GM/R^2.$$

From this equation one can calculate exactly how the radius R of our sphere changes with time. We shall, however, forgo such a calculation and try instead to analyse qualitatively the dependence of R on time.

First of all, we wish to emphasize the following important property of the acceleration written above. Expressing the mass M of the sphere in terms of the matter density ρ and the volume of the sphere $\frac{4}{3}\pi R^3$, we get

$$a = -\tfrac{4}{3}\pi \, G\rho R. \tag{10}$$

This formula tells us that the acceleration a is directly proportional to the distance. Thus, both the recession velocities of galaxies and their accelerations (decelerations) are proportional at present to their distances. But if at some moment both the velocity and its rate of change depend linearly on the distance, such a linear dependence will persist at a later time too. This pertains also to the preceding moments of time. As a result, the recession velocities of galaxies in the Friedmann model are always proportional to their distances, although the proportionality coefficient changes with time. The expansion decelerates, and at earlier times this coefficient was larger. One can plot a diagram showing how the radius of the sphere R changed with time in the past. The separation of any two distant galaxies in the Universe behaved in a similar way. Depending on whether the contemporary value of this separation is greater or smaller than R, one has respectively just to stretch or contract the ordinates of the plot for our selected sphere. Some such plots are shown in Fig. 7. The further in the past, the smaller was the radius of the sphere R (it is plotted with a solid line). The curve is bent in accordance with the deceleration of the expansion by the forces of gravity. Dashed lines in Fig. 7 show the past behaviour of distances to some other galaxies which at present are closer or farther than the surface of our sphere R. The ordinates of the dashed lines differ from those of the solid line by some fixed factors. The most important feature of this diagram is that at some moment in the past all the distances were zero. This was the moment of the beginning of the expansion. How long ago did it occur? How far apart are the points 0 and t_0 in Fig. 7? The answer depends on the present rate of expansion (on the Hubble constant H), i.e. on the tilt of the curve in Fig. 7 at the present epoch t_0, and on its curvature. The curvature is determined by the gravitational acceleration which, according to eq. (10), is determined by the density of matter in the Universe. If the expansion were not slowed down by gravity (assume for a moment that the matter density is vanishingly small and the deceleration a can be ignored), the galaxies would move apart by inertia with constant speeds. Instead of the curves of Fig. 7 one would obtain the straight lines shown in Fig. 8. In such a simple case the time t, that has elapsed since the onset of expansion, is determined by the value of the Hubble constant only, namely

$$t = 1/H \approx 1/75 \text{ km s}^{-1} \text{ Mpc}^{-1} = 13 \times 10^9 \text{ years.} \quad (11)$$

The range of uncertainty of H $50 \text{ km s}^{-1} \text{ Mpc}^{-1} < H < 100 \text{ km s}^{-1} \text{ Mpc}^{-1}$ leads to the following range of uncertainty of t:

$$10 \times 10^9 \text{ years} < t < 20 \times 10^9 \text{ years.} \quad (12)$$

The finite matter density in the Universe gives rise to gravitational forces which slow down the expansion and decrease somewhat the value of t (see the dotted curve in Fig. 8). Unfortunately, the mean matter density of the Universe has been determined with an unsatisfactorily low accuracy. The easiest to account for is the matter constituting the galaxies. The masses of individual galaxies can be estimated by motions of stars or other shining objects inside them. Having measured the velocities of such motions and the dimensions

Fig. 7. The time dependence of distances to remote galaxies. Different curves correspond to different galaxies; t_0 the present epoch, 0 the beginning of the expansion.

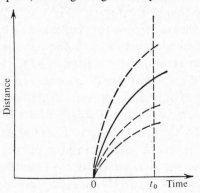

of a galaxy, one can calculate its mass from Newtonian laws of mechanics and gravity. Knowing galactic masses and the average number of galaxies per unit volume of space, one readily calculates the average density of cosmic material contributed by galaxies. This density averaged over the whole of space is about $\rho \approx 3 \times 10^{-31}$ g cm^{-3}. But intergalactic space may be filled with matter in such forms that practically neither emit nor absorb any light and, hence, are very difficult to detect. These might be, for example, a fully ionized intergalactic gas, weakly radiating, or completely extinguished stars. In addition, the Universe may contain a lot of neutrinos – pervasive particles which are extremely weakly coupled to ordinary matter and, as a result, are very difficult to register. Gravitational waves predicted by Einstein's theory may also contribute to the mass. There are other forms of matter between the galaxies too. It is very difficult to account for all these barely detectable kinds of matter. The most likely limits, within which the average density of all forms of matter seems to lie, are 5×10^{-29} g cm^{-3} – 3×10^{-31} g cm^{-3}. We discuss this problem in greater detail in §9. For a mean density within the above limits the gravity of the matter has a rather small effect on the value of t given in eq. (12). Thus, the expansion of the Universe began some 10–20 billion years ago. It is interesting that the age of the Earth, as determined from radioactive decay measurements, is 5×10^{9} years. Starting from this estimate, the Soviet physicists Ya.B. Zel'dovich and Ya.A. Smorodinsky obtained an upper boundary to the mean density of all possible elusive forms of matter in the Universe. The

Fig. 8. As for Fig. 7 but for a vanishing matter density in the Universe. For comparison the solid line of Fig. 7 is plotted in dots.

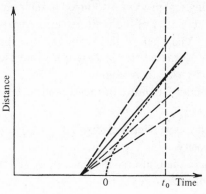

point is that the age of the Earth is undoubtedly less than the time that has elapsed since the beginning of the expansion of the Universe. If so, the maximum possible curvature of the curve in Fig. 8 corresponds to the time interval between the present epoch and the onset of expansion just equal to the age of the Earth. The curvature of this extreme curve tells one the value of the gravitational acceleration which, in its turn, enables one to calculate from eq. (10) the maximum possible matter density in the contemporary Universe. This maximum value is 2×10^{-28} g cm^{-3}.

It is interesting to compare the time t, as calculated above to have elapsed since the beginning of expansion, with the ages of other objects in the Universe. The typical age of the so-called globular star clusters in our galaxy is, for example, 10–14 billion years.

So, we see that the age of our own planet, as well as the ages of star clusters, are only slightly less than the value of t.

Let us return to the law of expansion of the Universe.

Thus, in the remote past, some 10–20 billion years ago, near the starting point of cosmological expansion the matter in the Universe was much denser than now. Individual galaxies, stars, and other objects could not exist as isolated bodies. All the matter was in the form of a continuous uniform medium. And only later in the course of subsequent expansion has it fragmented into separate clumps, which then evolved to form the variety of cosmic bodies observed now. We shall come back to this in Chapter 4.

Immediately, a number of questions arise. How reliable is the conclusion about the beginning of the expansion, the extremely dense phase of the whole of the Universe (or as astrophysicists put it – the *singular state*)? What physical processes occurred in such superdense matter? What forced the matter of the Universe into an expansion? And, finally, what was there before the beginning of expansion – before the singularity? All these problems are quite interesting and important, and we shall tackle them below.

First of all, let us discuss the future of the expanding Universe.

7   The future of the expanding Universe
The critical density

The expansion of the Universe is decelerating, and there are two possible versions of the future.

The rate of deceleration, as was shown in §6, is proportional to the density of matter in the Universe. As the expansion continues, the density decreases – and so does the rate of deceleration. It is possible that, for a given value of the present-day expansion rate, the matter density and the deceleration rate are small enough for the expansion to continue forever. Such a case is shown in Fig. 9a. The separation of any pair of galaxies (not bound gravitationally) would continue indefinitely.

Another possibility is that the density and thus the rate of deceleration of the expansion is sufficiently high for the expansion to slow down to a halt and be followed by a contraction. The effect on the separation of two galaxies for this case is shown in Fig. 9b.

The situation under discussion is completely analogous to that of a rocket boosted to a velocity which enables it to escape the planet. The velocity 12 km s^{-1}, for example, is quite enough for a rocket to leave the Earth and escape into space, since this velocity exceeds the 'escape

Fig. 9. (*a*) Variation of distance between two galaxies with time for a density of matter in the Universe less than the critical value. (*b*) The same dependence plotted for a matter density in excess of the critical value. The expansion of the Universe reverses to become a contraction.

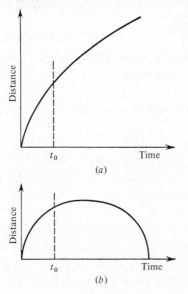

velocity' of the Earth. But such a speed would not be high enough to enable it to escape from the surface of Jupiter because the Jovian escape velocity is 61 km s^{-1}. A body launched from the surface of Jupiter with a speed of 12 km s^{-1}, having ascended to a certain altitude, would fall back to its surface.

Now, consider a galaxy at the surface of the sphere in Fig. 2. The velocity with which it recedes from the centre O is determined by the Hubble law, $v = HR$. If this velocity exceeds the escape velocity for a sphere of radius R, the galaxy will recede from O indefinitely and the Universe will expand forever (Fig. 9a); if, however, v is less than the escape value, the expansion will stop and reverse into a contraction (Fig. 9b). The velocity $v = HR$ is specified by the Hubble law, and which case – that of Fig. 9a or of Fig. 9b – accords with reality is determined by the mass of the sphere, i.e. depends on the matter density ρ.

Thus, with the present rate of expansion (for today's value of the Hubble constant, i.e. 75 km s^{-1} Mpc^{-1}) the Universe is destined to expand infinitely for a low matter density, while for a high density the expansion must stop and be followed by a contraction. Clearly, there exists a critical value ρ_{crit} of matter density separating these two cases.

It is not difficult to calculate this critical value of the density. Recalling a well-known formula for the escape velocity of a sphere with mass M,

$$v = (2GM/R)^{\frac{1}{2}}, \qquad (13)$$

and substituting for M and v the expressions $\rho\frac{4}{3}\pi R^3$ and HR respectively, we arrive at the equation

$$HR = (\tfrac{8}{3} G\pi\rho R^2)^{\frac{1}{2}},$$

which immediately gives the following expression for the critical density:

$$\rho_{crit} = 3H^2/8\pi G. \qquad (14)$$

Thus, the critical value of the mean density in the Universe is determined by the value of the Hubble constant H. With H equal to 75 km s^{-1} Mpc^{-1}, the critical density is

$$\rho_{crit} = 10^{-29} \text{ g cm}^{-3}. \qquad (15)$$

We see that the ultimate fate of the Universe depends on the actual value of the mean density of all kinds of matter in the Universe.

We have already mentioned in §6 that if one were to spread out uniformly all the matter locked up in visible galaxies, one would obtain the mean value $\rho_{gal} \approx 3 \times 10^{-31}$ g cm^{-3}, which is much less than the critical value ρ_{crit}. It is possible, however, that a great deal more barely detectable material resides between the galaxies. This question is of major significance. In the next sections we take a closer look at the main structural units of the Universe – the galaxies and the clusters of galaxies – and discuss in more detail the problem of intergalactic matter.

8   Galaxies and clusters of galaxies

Our survey of the properties of galaxies and their classification is, of necessity, rather brief. Much more comprehensive – though quite popular – reviews can be found in the following books edited by S.B. Pikelner: *The Origin and Evolution of Galaxies and Stars'*, Nauka, Moscow, 1976; *The Physics of the Cosmos*, Sovietskaya Entziklopedia, Moscow, 1976. Both of these books give references to a further special literature. To readers in the West we can recommend a popular book by S. Mitton, *Exploring the Galaxies*, Faber & Faber, London, 1976.

Galaxies are gigantic star systems each containing from a few million to a few hundred billion stars. Besides stars, galaxies also contain interstellar gas, interstellar dust and cosmic rays. The amount of gas inside the galaxies is not large compared to the mass of the stars, and varies from one galactic type to another. The amounts of matter in other forms are smaller still.

Four different types of galaxies are specified.

Elliptical galaxies comprise some 13% of all those that are comparatively close to us (brighter than 13th magnitude). (The classification of nearby galaxies into different types is presented according to G. de Vaucouleurs.) They are designated by the letter E and have a spherical or ellipsoidal shape. The spectra of these galaxies reveal that stars inside them move in all directions with almost equal probability, while the rotation of the galaxy as a whole is usually quite small. The density of stars per unit volume is highest in the centre and gradually decreases towards the edge. Most elliptical galaxies are very deficient in gas, which contributes less than 0.1% of the total mass. An example of a typical E galaxy is shown in Fig. 10.

Another class embraces spiral galaxies which are designated by the letter S. The proportion of spirals from among all the nearby galaxies is somewhat more than 60%. They are distinguished by the presence of two (and sometimes more) spiral arms making up a flat subsystem – the 'disk' (Fig. 11). Beside the disks, S galaxies possess so-called spherical components which are composed of objects scattered in a spherically symmetric manner around the centres of galaxies. The spiral arms contain a large number of bright young

Fig. 10. A galaxy cluster, Abell 1060, showing several galaxy types, including two giant elliptical galaxies in the centre of the field (Anglo–Australian Observatory).

stars and luminous gas clouds, heated up by these stars. Cold clouds of gas and dust are also present.

The dominant contribution to the mass of a spiral galaxy usually comes from the disk. In contrast to the spherical component, the stars

Fig. 11. M83, a spiral galaxy about eight million light years from our galaxy (Anglo–Australian Observatory).

and the gas in the disk rotate around the galactic centre with an angular velocity which varies with the distance from the centre.

The amount of gas in spirals ranges from 1 to 15% of their total mass.

The gas in galaxies (not only in spirals but in those of other types too) consists of 70% by mass of hydrogen, and 30% of helium. The admixture of other elements is quite small. The explanation of this fact will be given in Chapter 3.

The bulk of the galactic gas is in the form of neutral atoms. The temperature of the gas strongly depends on its density and a number of other factors. The gas is heated by soft cosmic rays and by ultraviolet and X-radiations. As a result, its temperature varies from 10 K in dense clouds to a few thousand degrees in a tenuous inter-cloud medium. It has been established recently that hydrogen in such dense cold clouds is predominantly in the molecular state. In the neighbourhoods of hot stars the gas is ionized by their ultraviolet emission, forming the so-called HII regions of ionized hydrogen (recall that hydrogen is the predominant element by mass). The mass of the ionized hydrogen around a single star may be as large as 10^4 solar masses (the mass of the Sun equals 2×10^{33} g and is designated $M_\odot$). The temperature in HII regions is about 10^4 K.

The next class of galaxies consists of lenticular galaxies designated SO. Of the nearby galaxies they account for about 22%. In these galaxies a bright flattened main body – the 'lens' – is surrounded by a weaker halo (Fig. 12). Sometimes the lens has a ring around it.

About 4% of all the nearby galaxies make a group of irregular galaxies. They are designated Ir. This class includes all those galaxies that cannot be assigned to one of the above three groups. The class of irregular galaxies is quite inhomogeneous. Fig. 13 presents an example of one such galaxy. The fraction of gas in the total mass of an irregular galaxy may sometimes reach 50%, but in other cases it may be as low as a few per cent.

The masses of galaxies vary over a very wide range, and so do their luminosities. The mass of a galaxy can be determined by the motion of stars and gas clouds inside it. In spirals, measurements of the spectral line shifts enable one to determine the rotational velocity as a function of distance from the centre. Then, using the law of gravity, it is a straightforward procedure to calculate the mass of the galaxy.

For ellipticals, which do not show noticeable rotation, the masses can be estimated from the dispersions (scatter) of stellar velocities. The velocity dispersion results in the broadening of spectral lines. The larger the velocity dispersion – the larger the spectral line width – the

Fig. 13. Galaxy M82 in Ursa Major, an irregular object dominated by dust (S. Mitton).

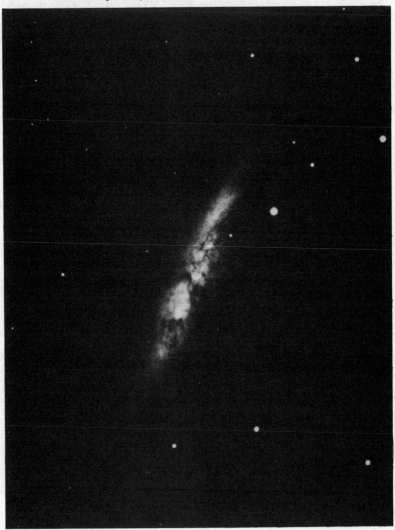

more massive is the galaxy. The most diverse are the elliptical galaxies. There are supergiants among them which emit tens of times more light than our galaxy does, and have masses up to 10^{13} $M_\odot$ (the mass of the Galaxy is some 10^{11} $M_\odot$). But there are also dwarf ellipticals – the pygmies – whose radiative power is tens of thousands of times smaller than that of our galaxy, while their masses amount to a meagre 10^6 $M_\odot$. Supergiant ellipticals, however, are encountered rather rarely. Irregular galaxies usually have comparatively low luminosities (0.1–0.01 of the luminosity of the Galaxy) and not particularly large masses (10^{10}–10^8 $M_\odot$).

Some galaxies are powerful emitters of radio waves; their power at radio frequencies by far exceeds that in the optical range. These galaxies are called radio-galaxies.

The majority of radio-galaxies emit most of their radio power from extended regions (hundreds of thousands of parsecs in size) disposed symmetrically on either side of the optically emitting galaxy (Fig. 14).

The central regions of many luminous galaxies have bright concentrations called the nuclei, and inside the nuclei one can sometimes discern brilliant points. The physical nature of galactic nuclei differs dramatically from that of their other regions. Violent processes occur there, accompanied by the liberation of vast amounts of energy. The importance of these processes had been stressed in 1958 by V.A. Ambartzumyan.

Some galaxies have unusually active nuclei. Such are the so-called Seyfert galaxies, N-galaxies and others. In their nuclei violent motions of gas with velocities of thousands of kilometres per second are observed. Large amounts of matter are frequently being ejected; the brightness of the nucleus often displays strong variations.

A unique group of extragalactic objects is represented by quasars discovered in 1963 by the Dutch astronomer (working in the USA) M. Schmidt. These objects emit hundreds of times more optical light than galaxies, and the main part of it comes from a compact nucleus of no more than 0.1 pc in size! The quasar nuclei are surrounded by gas envelopes extending to hundreds of parsecs. The quasars are powerful sources of radio emission, and some of them, in addition, of infrared radiation and X-rays. The optical brightness of quasars is variable, resembling in this respect the optical brightness of active galactic nuclei.

Two years after the discovery of quasars A. Sandage found the so-called quasi-stellar galaxies, which are almost the same as quasars except for a much fainter radio emission.

Now it is widely believed that quasars are the nuclei of galaxies in a state of extremely violent activity. The stars of the galaxy surround-

Fig. 14. Cygnus A, a giant radio galaxy and one of the most energetic galaxies known (Hale Observatories).

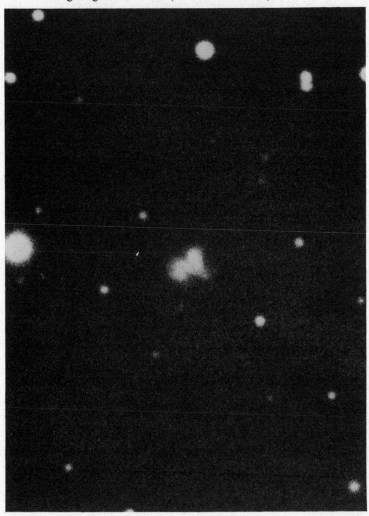

ing the quasar normally cannot be seen, because at such immense distances the light of the bright quasar completely swamps a weak stellar emission.

In recent years weak extended optical halos have been discovered around a few tens of close quasars. Typical sizes of these halos are about 90 000 pc, while their optical luminosities fall a few times below the luminosities of brightest galaxies. They were found to contain large amounts of ionized gas – mostly hydrogen – but it is not clear so far whether they contain stars.

The majority of all galaxies are clumped into clusters. Catalogues have been published which contain thousands of clusters of galaxies. Among all the clusters one may distinguish between regular and irregular ones. Apart from this important division into two groups, a number of classifications exist that distinguish clusters by some other parameter, for example by richness (the number of members brighter than some limit), by the presence of bright galaxies in the central regions, by the presence of peculiar galaxies and so on. Regular clusters consist of a large number of individual galaxies (sometimes more than 10^4) forming a spherical system with a high degree of concentration towards the centre. The bright members of such clusters seem to belong to E and SO types only. The very centre of the cluster is often marked by the presence of one or two brightest elliptical galaxies surrounded by haloes. These galaxies are called CD galaxies. As a typical example of a regular cluster one can point out the cluster of galaxies in Coma. Irregular (dispersed) clusters have irregular shapes and often include isolated condensations. These clusters are composed of galaxies of all types. They may be rich (with more than a thousand members) or poor. The Soviet astrophysicist B.A. Vorontsov-Velyaminov maintains that the galactic field in general consists of loose outer parts of numerous overlapping dispersed clusters plus small groups of galaxies.

The best studied are the regular clusters. The size of the regular cluster in Coma is about 4 Mpc. The total number of galaxies in it (including the faintest) is estimated as approaching a few tens of thousands. The dispersion of galactic radial velocities is $\Delta v \approx 1000$ km s^{-1}.

The Virgo cluster of galaxies provides an example of an irregular cluster. It has thousands of members and its size is about 3 Mpc.

In some clusters of galaxies large amounts of hot ionized gas with temperatures of about 10^8 K have been discovered. Such gas is a strong emitter of X-rays. The total mass of the hot gas in a cluster may comprise an appreciable fraction of the total mass of all the galaxies. Interaction of this hot gas with the relict radiation permeating the whole of the Universe (we discuss this in more detail in §4 of Chapter 3) results in an interesting observational effect predicted by the Soviet astrophysicists Ya.B. Zel'dovich and R.A. Sunyaev. This effect appears as a slight diminution in the intensity of the relict radiation at centimetre wavelengths caused by scattering of relict photons off the hot plasma electrons inside the cluster. Thus, when observing the relict radiation through a cluster of galaxies, one should register somewhat lower flux values as compared to the neighbouring directions away from the cluster. Combining these observations with X-ray data for the same cluster, one can determine linear dimensions of the hot gas cloud in the cluster. With its angular size known, one then readily calculates the distance to the cluster. Thus, the Zel'dovich and Sunyaev effect may, in principle, serve as an additional method of determining the distances of clusters of galaxies and, as a consequence, of the value of the Hubble constant.

The possible existence of clusters of clusters galaxies has been disputed for a long time. It seems to have been established now that superclusters really do exist. But apparently, no pronounced density inhomogeneities occur on scales tens of times larger than the dimensions of big clusters. The most compelling evidence for this came indirectly, through the observational constraints on the anisotropy of the relict radiation – a fact already mentioned above and to be discussed in more detail in Chapter 3. According to the American astrophysicist F. Zwicky the diameters of the biggest clusters are about 8 Mpc. The size of the largest space region over which a still significant density contrast occurs is about 40–100 Mpc, corresponding to the mean separation of big clusters. On scales larger than 100–200 Mpc the Universe is homogeneous.

In addition to big clusters, there are a great many small clusters, groups and pairs of galaxies.

Recently, the analysis performed by the Soviet astrophysicists J. Einasto, E. Saar, M. Joeveer and others and by the American scientists P.J.E. Peebles, S.A. Gregory, L.A. Thompson and others has revealed that the largest inhomogeneities in the spatial distribution of galaxies have a 'cellular' structure, the 'walls of the cells' being the regions of enhanced galaxy density with the 'void' between them. The typical size of these cells is about 100 Mpc, the thickness of their 'walls' is some 3–5 Mpc. Big clusters of galaxies usually reside at the 'junction points' of this structure. Individual fragments of the cellular structure are sometimes called superclusters. The exploration of these important issues is currently under way.

9 The mean density of matter in the Universe and the problem of the 'hidden' mass

Now let us turn back to the problem of the mean density of matter in the Universe. As has been mentioned already, one can, without much difficulty, take into account 'easily detectable matter', i.e. the matter constituting visible galaxies. A quite reliable estimate of their contribution was performed in 1958 by the Dutch astronomer J. Oort. In practice, it takes two steps to determine the average density of matter contained in galaxies.

First of all, one counts the galaxies of various luminosities in a large enough volume of space and calculates the average luminosity per unit volume of the Universe. For this Oort obtained

$$\mathscr{L} = 2.2 \times 10^{-10} \, L_\odot/\text{pc}^3. \tag{16}$$

Here $L_\odot$ is the luminosity of the Sun: $L_\odot = 4 \times 10^{33}$ erg s^{-1}.

After that, one calculates for all types of galaxies the ratios of the masses M to their luminosities L. The average mass–luminosity ratio M/L, obtained in this way for elliptical galaxies, is 50 times that of the Sun, $M_\odot/L_\odot$. For spirals, the ratio M/L varies from few times $M_\odot/L_\odot$ to about 20 $M_\odot/L_\odot$. Allowing for the percentage of galaxies of every galactic type, one arrives at the following average value

$$M/L \approx 21 M_\odot/L_\odot. \tag{17}$$

The product of eq. (16) and eq. (17) gives the mean density of matter condensed into galaxies:

$$\rho_{\text{gal}} \approx 4.6 \times 10^{-9} \, M_\odot \, \text{pc}^{-3} = 3 \times 10^{-31} \, \text{g cm}^{-3}. \tag{18}$$

This value is significantly less than the critical density given in eq. (15). The ratio of these two density values, usually denoted by the letter Ω, is

$$\Omega_{\text{gal}} \equiv \rho_{\text{gal}}/\rho_{\text{crit}} \approx 0.03. \tag{19}$$

If the Universe does not contain an appreciable amount of some other kind of matter with an average density considerably exceeding ρ_{gal}, it will expand forever.

There are serious grounds for suspecting that much matter is concealed between the galaxies; this matter – called the 'hidden mass', or the 'missing mass' – could be in forms that are very difficult to detect with conventional methods.

One of the motives for such a suspicion stems from the measurements of masses of clusters of galaxies. These measurements are performed as follows.

Regular clusters have a symmetric shape; the density of galaxies in them gradually decreases from the centre to the outskirts. Hence, it is quite reasonable to assume that such clusters are in a state of equilibrium, in which the kinetic energy of the motion of the galaxies is balanced by the mutual gravitational attraction of all bodies in the cluster.

In this case we may apply the virial theorem, which states that the time-averaged total kinetic energy of all cluster members is half the absolute magnitude of the total potential energy of the gravitational interaction of the total mass (including the invisible mass) in the cluster. This theorem makes it possible to estimate the total mass of the cluster, once its size and the relative velocities of the galaxies are measured. The relative velocities of the galaxies are inferred from their red shifts, while the size of the cluster can be readily calculated from its angular diameter and distance. Such a procedure performed for the above mentioned Coma cluster results in a mass estimate of some $2 \times 10^{15} \ M_\odot$, which results in the mass–luminosity ratio of the cluster as a whole (according to G.O. Abell):

$$M/L \approx 150 \ M_\odot/L_\odot.$$

The above value is much too high even for the elliptical galaxies, which have the greatest mass–luminosity ratios (these data are being revised, however). The ensuing conclusion is that the actual mass of the Coma cluster is much greater than the sum of masses of the individual galaxies belonging to it. Similar results have been obtained for other clusters and groups of galaxies too. In this way the problem of 'hidden mass' arose. We note at once, however, that to infer the mass of a cluster of galaxies from the virial theorem is not a simple task and considerable errors may be introduced. The main source of errors is the uncertainty in the measured values of galaxy velocities: having, say, overestimated the velocity dispersion, one exaggerates the mass of the cluster. In addition, chance projection of 'alien' galaxies is always possible, which also adds to the mass of the cluster. Nonetheless, a more thorough analysis shows that it is very difficult to put all the blame for the paradoxically large values of cluster masses on errors of that kind alone. As a result, the above conclusion gives a strong impetus to serious efforts in the search for a 'hidden mass' – and not only in clusters of galaxies but between them too. In what forms of matter could 'the hidden mass' be hidden? Perhaps in the form of intergalactic gas?[†] Obviously, the intergalactic space has much greater volume than the space occupied by galaxies. As a consequence, the intergalactic gas, though much more tenuous than the gas inside the galaxies, could in principle dominate in terms of mass.

First of all recall that the main constituent of the gas in the Universe is hydrogen. Hence, in order to find the intergalactic gas, one has in the first place to look for hydrogen. Depending on physical conditions, the gas can be in either the neutral or ionized state.

Let us begin with an estimate of the possible amount of neutral hydrogen.

When light from a remote source passes through gas containing neutral hydrogen atoms, it is absorbed (or rather resonantly scattered) at a wavelength $\lambda = 1216$ Å of the L_α spectral line. As a result, the light of the source is attentuated at this particular wavelength. In practice, distance quasars are used as the light sources. Hydrogen atoms are spread over the whole distance from a quasar to the Earth and, consequently, have different recession velocities due to the expansion of the Universe described by the Hubble law $v = HR$. Doppler shifts due to different relative velocities cause a narrow absorption line to broaden to a wide absorption band in the spectrum of a quasar. The spectra of a number of quasars with red shifts $z > 2$ have been thoroughly searched for such an absorption band, but nothing has been found. From this a conclusion can be drawn that the mean number density of neutral hydrogen atoms in

[†] Many astrophysicists have contributed to the analysis of observational data concerning intergalactic gas. The following Soviet theoreticians, among others, may be mentioned: V.I. Ginsburg, Ya.B. Zel'dovich, I.S. Shklovsky, A.G. Doroshkevich, V.G. Kurt, L.M. Ozernoi and R.A. Sunyaev.

intergalactic space is negligibly small, namely $n_{HI} < 10^{-11}$ cm^{-3}, which corresponds to a matter density

$$\rho_{HI} < 10^{-35} \text{ g cm}^{-3}.$$

The same argument applies to molecular hydrogen (to the absorption in the Lyman band of hydrogen molecules) too. The observations tell us that the density of intergalactic hydrogen in the molecular state is also negligible.

Thus, the intergalactic gas – if it exists – must be ionized and, consequently, very hot. As a more detailed analysis shows, its temperature must be in excess of a million degrees. One should not be surprised, however, that despite such a high temperature the gas is practically invisible. The fact is that, being extremely rarefied, it is completely transparent and emits a negligible amount of visible light. But such a hot completely ionized plasma quite noticeably radiates in the ultraviolet and soft X-ray regions of the spectrum.

One possible way to try to detect the hot intergalactic gas is to look for its ultraviolet emission. But this method has not proved to be particularly sensitive.

Another interesting way has been proposed by the Soviet astrophysicist R.A. Sunyaev. It is based on the following idea. The ultraviolet radiative flux from the hot intergalactic gas must ionize hydrogen atoms at the peripheries of galaxies. On the other hand, the observations at radio frequencies have revealed the presence of neutral hydrogen at the outskirts of our galaxy and others. The calculations show that, if the mean density of the hot intergalactic gas were equal to the critical value $\rho_{HII} = 10^{-29}$ g cm^{-3}, the flux of the ultraviolet radiation would be strong enough to completely ionize the hydrogen at the peripheries of galaxies – in contrast to the observations. Hence

$$\rho_{HII} \lesssim 10^{-29} \text{ g cm}^{-3}.$$

This value is much more than ρ_{gal}. Thus, unfortunately, the method under discussion is not sensitive enough to exclude the possible existence of large amounts of hot gas between the galaxies. The question of a reliable estimate for the mean density of such a gas – whether it exceeds the average density of galactic matter or not – remains open.

Let us turn now to the gas in clusters of galaxies. Radio observations indicate that the amount of neutral hydrogen in clusters is extremely small. But X-ray telescopes mounted on satellites have successfully detected the emission of a hot ionized gas in rich clusters of galaxies. This gas turned out to be heated up to $T \approx 10^8$ K. Its total mass can be as high as 10^{13} $M_\odot$. Nevertheless, no matter how impressive the number quoted might look, it cannot match the above mentioned total mass of the Coma cluster, 10^{15} $M_\odot$, deduced from the virial theorem. Thus, the discovery of the hot gas in clusters of galaxies has not resolved the problem of hidden mass.

A few years ago this notorious problem acquired a new aspect.

Recently there have appeared more and more advocates of the idea that the galaxies are surrounded by extended massive haloes, consisting of weakly radiating objects that easily elude detection. These objects could, for instance, be low-luminosity stars. The gravity of such halo stars would have little effect on the dynamics of the internal regions of galaxies as they are observed (recall that a spherical shell creates no gravitational field in its interior (see § 2)); as a result, the observations of internal galactic regions can reveal the masses of the interiors themselves but not those of the haloes. Massive haloes, however, must influence the motions of dwarf galaxies orbiting the main galaxy. It is this influence that has been taken as the basis for the attempts now being made to disclose the haloes of galaxies. It is quite possible that the haloes of galaxies, when properly accounted for, will change appreciably the values of galactic masses and settle the problem of hidden mass. At present, however, the issue has not yet been resolved.

Finally, we have to discuss more exotic possible contributors to the hidden mass – such as cosmic rays, neutrinos, gravitational waves and other physical forms of matter.

Observational data provide good evidence that the mass density of cosmic rays does not exceed 10^{-35} g cm^{-3}, i.e. they can be ignored.

The situation becomes more complicated when one considers neutrinos and gravitational waves. The matter in these forms interacts extremely weakly with ordinary substances and, therefore, even if the Universe were filled with neutrinos or gravitational waves of a mass density (related to the energy density through Einstein's formula $\varepsilon = \rho c^2$ in the case of relativistic particles) in excess of ρ_{crit}, direct physical methods would be of no use in trying to detect them. There are, though, some indirect arguments against the large mass density of relativistic particles in the Universe (see § 7 of Chapter 3).

Quite recently the results of a few physical experiments have been reported which claim a non-zero rest mass for neutrinos. If they turn out to be correct, then the Universe should contain a considerable amount (by mass) of this exotic kind of matter. For more details on this see Chapter 4.

Summing up the above discussion, we conclude that the problem of the mean matter density ρ in the Universe has not been resolved yet. We shall return to this issue once more in § 4 of Chapter 2, where we discuss a method of determining ρ which does not depend on the specific physical nature of matter but rather makes use of the ability of mass in any form to create gravitational fields. But actually even this universal method has had little success as yet.

In conclusion, it is worth noting that the majority of researchers in the field believe that the most likely value for the mean density of all kinds of matter in the Universe, as derived from all existing observational data, is

$$\rho \approx 0.1 \ \rho_{crit}. \tag{20}$$

Though the truth in science is not established by a vote among specialists, the reader will probably be interested to know that these specialists believe that the matter density in the Universe is below the critical value and the Universe will expand forever. This conclusion may change, however, if the neutrinos are proved to have non-zero masses, which are not all too small (see Chapter 4).

10   Does the red shift prove the expansion of the Universe?

As we have already mentioned in the Introduction, the idea of a non-stationary Universe was not immediately accepted by all. The principal importance of this far reaching conclusion aroused innumerable objections based on prejudice. The grandeur of the ensuing picture of the expanding Universe was too stunning. The concept of a non-evolutionary static Universe seemed much more natural and simple. This inspired numerous attempts to uphold the idea of a stationary Universe, to find some alternative explanation for the 'cosmological red shift', other than the Doppler shift caused by recessional motions of galaxies. One could forgo any discussion of these attempts and leave them to historians of cosmology, had they been made in the past only, during the first years after the discovery of the red shift. But, regrettably enough, such attempts are sometimes being made even today. For this reason we are bound to raise the question: is the Doppler-effect interpretation of the red shifts of spectral lines in galactic spectra the only one possible? Couldn't some other physical effect cause the reddening of light quanta – the photons?

General relativity theory establishes that the light quanta shift toward the red when they propagate from regions with large absolute values of gravitational potential to those of smaller potentials, i.e. come out of regions with strong gravitational fields. For example, the photons leaving the surface of the Sun are reddened, and so are the photons propagating upwards in a terrestrial laboratory. On the other hand, photons moving downwards in a terrestrial experiment are shifted toward the blue. These effects, though extremely weak, have been successfully measured.

But there is no way in which the effect of the reddening of photons coming out of strong gravitational fields could account for the cosmological red shift of galactic light. Firstly, it would be too weak in a homogeneous Universe with its current value of the mean matter density; secondly, a spectral shift of this origin is proportional to the square of distance not directly proportional to it; thirdly, – and most important – it would in fact be a blue shift instead of a red one! This last point is easy to understand. To do so let us return once more to the imaginary sphere of Fig. 2. Suppose that the light is emitted by galaxy A at the surface of our sphere and propagates toward the observer at its centre O. Recall that the matter outside the sphere has no gravitational effect whatsoever on its interior, and one can consider light photons as propagating in the gravitational field of the sphere itself only. The motion of a photon from the surface of the sphere toward its centre is analogous to the downward motion of photons on the Earth, i.e. it results in the blue shift.

Of course, the gravitational frequency change of photons is accounted for in exact formulae of the relativistic theory of the cosmological red shift, but this is a 'second-order effect'. It never dominates in the total value of the red shift. We did not mention it before, having postponed the discussion of all the relativistic effects for subsequent chapters since we are considering the limit of small distances only, where relativistic corrections can be safely ignored.

One of the ideas proposed to explain the cosmological red shift appeals to a hypothetical energy loss by photons while passing through vast extensions of space. Since the photon energy is proportional to its frequency in accordance with the formula $E = h\nu$, where $h = 6.65 \times 10^{-27}$ g cm^2 s^{-1} is the Planck constant, a decrease in energy implies a decrease in frequency or, in other words, an increase in wavelength – the reddening of a photon.

A possible mechanism for such an energy loss could be due to the interaction with intergalactic matter (or with radiation in some versions of the hypothesis) while travelling for a long time through intergalactic space. In this case, however, the encounters with particles or photons would be accompanied not only by a decrease in the energy of the photons, but also by a change in their direction of motion. In this respect such an interaction would be similar to the scattering process. Such an effect would result not only in the reddening of quanta, but in the smearing of galactic images also. Nothing of the kind has been observed. So, the concept has to be rejected. There is no need, in fact, to dwell on other difficulties of this hypothesis.

Another conceivable mechanism of reddening could be the hypothetical spontaneous decay of light quanta accompanied by the emission of other particles.

Even as early as 1934, the Soviet physicist M.P. Bronshtein put forward an argument which led to a conclusion that, were such a decay process to exist, it would have long since been detected in laboratory experiments. The point is that the probability of such hypothetical spontaneous decay would be inversely proportional to the photon frequency and, as a consequence, the observed red shift would be different for different frequencies – which is not the case either for the observed galactic spectra or for the Doppler effect itself. Bronshtein's argument proceeds as follows.

Consider how the lifetime of some unstable particle (a μ-meson, for instance) changes with an increase in its kinetic energy. Let T_0 be the lifetime of this same particle at rest. Then, according to special relativity theory, the lifetime of a particle moving with velocity v becomes

$$T = T_0/(1 - v^2/c^2)^{\frac{1}{2}}. \qquad (21)$$

The energy of the moving particle, having a rest mass m_0, is

$$E = m_0 c^2 / (1 - v^2/c^2)^{\frac{1}{2}}. \tag{22}$$

The ratio of T to E is constant:

$$T/E = T_0/m_0 c^2 = \text{constant}, \tag{23}$$

$$T = \text{constant} \times E.$$

The probability of decay W is inversely proportional to the lifetime T and hence, inversely proportional to the energy of the particle,

$$W = 1/T = \text{constant}/E. \tag{24}$$

Formula (24) is universal and can be applied to all particles including photons, although eqs. (21) and (22) are not directly applicable to photons moving with exactly the speed of light c. Now one has only to substitute $E = h\nu$ for the energy in eq. (24) to arrive at the conclusion that the probability of decay of a photon must be inversely proportional to its frequency. But if it were true, the fastest to decay would be the radio quanta. However, direct observations of the radio line $\lambda = 21$ cm for 30 remote galaxies carried out in the 1960s clearly demonstrated that the red shift at radio frequencies is exactly the same as in the optical.

As a result, the hypothesis of aging (or tiring) photons must also be discarded.

The Doppler shift caused by the expansion of the Universe remains as the only viable explanation of the cosmological red shift. (At very large distances one must account also for the effects of general relativity; see Chapter 2.)

We wish to emphasize once more that the non-stationary behaviour of the Universe had been predicted theoretically before its experimental discovery (see §2), and the measurements of the red shift have only borne it out. One is led to wonder not at the facts of the red shift and the expanding Universe (its non-stationary behaviour is a direct consequence of the laws of physics) but at the amazing tenacity of conservative views.

Had the observations revealed no systematic displacement of the spectral lines of distant galaxies, i.e. no evidence for the non-stationary character of the Universe, it would only mean that the laws of gravity (both Newton's and Einstein's) needed some revision, that universal forces of some unknown kind were preventing the forces of gravity from making the Universe non-stationary.

Curiously, an attempt to introduce such a new force had been made by Albert Einstein himself in the earliest days of contemporary cosmology, before the discoveries of Friedmann and Hubble. This is the subject of the next section. We need add only that some astrophysicists today discuss the possibility that, apart from the cosmological red shift due to the expansion of the Universe, a number of quasars and galaxies might possess extra red shifts caused by other physical effects, such as the effect of a strong gravitational field or some other as yet unknown phenomena. In principle, this is certainly possible. But the author believes that the observational data quoted in support of such ideas provide no compelling evidence and can be explained in other, more natural ways. For those interested in greater details we recommend the book *Confrontation of Cosmological Theories with Observational Data*, edited by M. Longair, Reidel, Dordrecht, 1974.

11   Is the vacuum gravitating?

The story of the scientific idea of a gravitating void, which we put forward in this section, begins with the same conflict between the traditional belief in the unchanging Universe and the concept of an evolutionary Universe inexorably ensuing from the laws of gravity.

The universal law of gravity asserts that any two material bodies are attracted to each other. But is the vacuum gravitating? This question was raised in modern physics by Albert Einstein in 1917. What is the gravitation of a vacuum? Why did such a question arise? What experimental data or astronomical observations forced the great physicist to raise this question? The fact is that no direct evidence existed or, more precisely, just the absence of data on the galactic motions directed Einstein's mind toward the idea of a gravitating vacuum.

Things were as follows. Soon after the completion of the theory of general relativity, Einstein made an attempt to construct on its basis a mathematical model of the Universe. This was before the works of Friedmann, before the discovery by Hubble of the red shift in the spectra of galaxies, and Einstein was led by the idea of a stationary, permanent Universe: 'The heavens endure from everlasting to everlasting'. But, as we have seen in § 2, the law of gravity requires a non-stationary Universe. The gravity of the uniformly distributed matter in the Universe produces a negative acceleration, given by eq. (10), which increases in direct proportion to the distance.

To balance the forces of gravity and make the Universe stationary, one has to introduce forces of repulsion independent of the specific properties of matter. The gravitational acceleration and the acceleration due to the repulsion must be equal in magnitude and opposite in sign:

$$a_{rep} = -a_{grav} = \tfrac{4}{3}\pi\, G\rho R. \tag{25}$$

The second part of the above equation is obtained by substituting the value for a_{grav} from eq. (10). Thus, the force of repulsion must be directly proportional to the distance.

Proceeding from such arguments, Einstein introduced a cosmic force of repulsion which made the Universe stationary. This force is universal: it does not depend on the masses of bodies, but on their separations only. The relative acceleration imparted by this force to any pair of bodies separated by a distance R must be proportional to R and, hence, can be written as

$$a_{rep} = \text{constant} \times R. \tag{26}$$

Knowing the mean matter density in the Universe (we assume it to be equal approximately to the critical value $\rho \approx 10^{-29}$ g cm^{-3}), one can estimate numerically from eqs. (25) and (26) the magnitude of the repulsion acceleration:

$$a_{rep} \approx 3 \times 10^{-36}\, R \text{ cm s}^{-2}. \tag{27}$$

The numerical factor in eq. (27) or, more precisely, the ratio of this factor to the square of the speed of light is called the cosmological constant and designated by Λ. Thus, according to Einstein,

$$\Lambda = 3 \times 10^{-36}/(3 \times 10^{10})^2 = 3 \times 10^{-57} \text{ cm}^{-2}. \tag{28}$$

In principle, the repulsive force – if it exists, of course, – could be discovered in precise enough laboratory experiments. But the extremely small value of Λ makes it a hopeless task. Indeed, one can easily evaluate that a body in a free fall near the surface of the Earth acquires a deceleration due to the repulsive force which is 31 orders of magnitude (!) less than the free fall acceleration itself. Even on the scales of the Solar System or of the Galaxy as a whole the repulsive forces are negligibly small compared to gravity. For example, one readily estimates that the acceleration of the Earth in the gravitation field of the Sun is 0.5 cm s^{-2}, while the acceleration of cosmic repulsion of the Earth from the Sun is, according to eq. (27), 10^{23}

times less! Clearly, such a repulsion has no effect whatsoever on the motion of cosmic bodies in the Solar System and can be detected only when exploring the motions of the most distant galaxies that can be observed.

This was how a cosmological term representing the vacuum repulsive force entered Einstein's gravity equations. The action of this force is as universal as that of gravity itself: it does not depend on the physical nature of interacting bodies, and it would be logically consistent to call it the gravity of a vacuum.

A few years after Einstein's paper appeared A.A. Friedmann established his famous laws. Having recognized Friedmann's contribution, Einstein was inclined to think that the Λ-term could perhaps be abandoned, once it was possible to find a solution to the field equations without this term, describing the Universe as a whole.

But after the red shift of galactic spectra appertaining to the expansion of the Universe had been discovered, all the motivation in assuming that some cosmic repulsive force is inherent in nature disappeared. It is true, however, that the solution describing an expanding Universe can be obtained from the equations with the Λ-term too. For this one has to assume only that the forces of gravity and repulsion do not cancel each other exactly, and then the prevailing force will ensure a non-stationary behaviour. This had been noted by Friedmann himself in his pioneering works. The red shift data of Hubble's time were not accurate enough to determine which solution occurs in reality – the one with the Λ-term or the one without it. Nevertheless, many physicists bore a grudge against the Λ-term in Einstein's equations because it complicated the theory and had no basis to it. Einstein himself and many other physicists preferred to operate with equations that had no Λ-term, i.e. to set $\Lambda \equiv 0$. Einstein called the introduction of the cosmological constant into his original equations 'the biggest blunder' of his life.

But the cosmologists of the 1930s were not quick to abandon the Λ-term completely. They had serious grounds for retaining it in the field equations. Recall that the first measurements of the Hubble constant obtained the value $H \approx 500$ km s^{-1} Mpc^{-1}, which implied that the start of expansion of the Universe occurred some $t \approx 1/H \approx 2 \times 10^9$ years ago (see §§ 5 and 6). This was too short a time. First, it turned out to be even less than the age of the Earth; second and more important, the age of stars and stellar systems was then erroneously estimated to be 10^{13} years, i.e. more than three orders of magnitude in excess of the expansion time given by Hubble.

We know today that the value of $t = H^{-1}$ was underestimated by a factor of 10 (see § 5), while the age of the stars, on the contrary, was overestimated by more than two orders of magnitude. And from today's point of view these two ages are in a reasonable agreement. But in the 1930s the above mentioned discrepancy was considered as a serious contradiction.

To reconcile the expansion time of the Universe with the ages of the stars, the Λ-term was employed again. Thus, the idea of the universal cosmic repulsion entered its period of 'second youth'.

Let us now try to understand how the introduction of the Λ-term can significantly change the expansion time of the Universe.[†]

Suppose that the Λ-term is not zero. Let the Universe expand from a state of very high density. Since the matter density at the beginning is high, the gravitational force slowing down the expansion, which is proportional to the density, completely dominates the repulsive force. Indeed, the ratio of the acceleration due to the gravity of matter to that generated by the cosmic repulsion is, according to eqs. (10) and (26),

$$a_{\text{grav}}/-a_{\text{rep}} = \tfrac{4}{3}\pi G \rho R/\text{constant} \times R = \text{constant}' \times \rho. \qquad (29)$$

(We do not assume here that $a_{\text{rep}} = -a_{\text{grav}}$, as was the case for Einstein's cosmological model. In the example being discussed the forces do not cancel each other out, the Universe is expanding and the density of matter drops.) At the initial phase of expansion, when the density ρ is large, the above ratio is also large, and the gravity force greatly exceeds the force of repulsion. As the expansion proceeds further, the density sooner or later drops to such a low value that the forces become equal. At this moment of equilibrium the Universe expands by inertia with a constant rate, being neither accelerated nor decelerated. Now, if we arrive at this equilibrium state with a low enough rate of expansion, the force of gravity could remain almost

---

† Note that the cosmological models with the Λ-term have been extensively studied by the Belgian scientist G. Lemaître, to whom unfairly the credit for the creation of the whole of the theory of the expanding Universe is sometimes ascribed in Western literature. As already emphasized, such a theory was first developed by A.A. Friedmann. Concerning the theory of the expanding Universe, Einstein wrote in 1931: 'The first to take this path was Friedmann'.

exactly balanced by the force of repulsion for a sufficiently long period such that this state of suspended expansion could be quite prolonged. Later, the matter density will eventually fall such that the force of repulsion will outweigh the gravity force. At this stage the expansion of the Universe will be accelerated by the repulsive force. The variation of distances with time in such a Universe with a delayed expansion is shown in Fig. 15. With a suitable choice of parameters of the model, one can make the delay in the expansion arbitrarily long. According to such a hypothesis the delay in the expansion must have occurred in the remote past. Today the expansion of the Universe is accelerating, and the contemporary value of the Hubble constant does not indicate the time that has elapsed from the start of the expansion. This is clearly demonstrated in Fig. 15.

This is how the introduction of the Λ-term enabled one to prolong the expansion of the Universe and reconcile it with the ages of the stars.

The value of the Hubble constant was revised in the 1950s. Earlier still, in the late 1930s, it had been established that the conversion of hydrogen into helium serves as the energy source in stars, and a modern theory of stellar evolution was developed in the 1950s. The

Fig. 15. The variation of distance with time between two galaxies for $\Lambda \neq 0$. At first the force of gravity slows down the expansion; then the force of gravity decreases and approaches the force of repulsion in magnitude – the expansion is suspended for a time; eventually, the repulsion 'wins' and the expansion speeds up. The dashed line shows how the expansion would proceed with today's value of the Hubble constant, were there no gravity and no repulsion. The time t deduced from the value of the Hubble constant bears no relation whatsoever to the actual expansion time of the Universe.

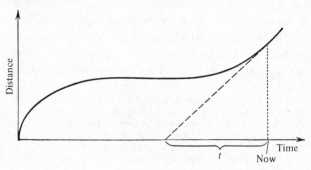

controversy over ages was resolved and the Λ-term became unnecessary for the second time already!

But in 1967 a period of 'third youth' began for the idea of the Λ-term. By this time astronomers had discovered and explored to some extent the most unusual objects – quasars, which have already been discussed briefly in § 8.

Even today quasars pose a number of mysteries and unsolved problems. Here, we touch only on two peculiar properties of quasars. As mentioned in § 5, the galaxies had long ago been shown to have apparent stellar magnitudes closely correlated with their red shifts (the m–z relation). The farther away the galaxy, the less bright it appears to the observer (the greater its stellar magnitude m), and the greater – according to the Hubble law – its velocity and red shift. It was found that quasars did not display such a clear correlation (see Fig. 16). The American scientists V. Petrosian, E. Salpeter and P. Szekeres suggested that the absence of the m–z relation for quasars is due to the Λ-term in gravity equations, or, in other words, due to the cosmic repulsion. Let us clarify this statement. The above mentioned authors argued that quasars were observed to lie at enormous distances, much farther away than the most distant galaxies accessible to telescopes. When we observe distant quasars with large red

Fig. 16. Apparent stellar magnitude versus red shift (multiplied by the speed of light in km s^{-1}) for quasars. The straight line is redrawn from Fig. 5.

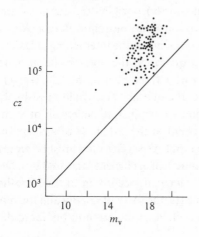

shifts, we are registering light emitted long ago. If this light had been emitted at the epoch of suspended expansion of the Universe governed by the equations with a Λ-term (see above), then quasars at greatly different distances can have almost the same red shifts. This occurs because we are observing cosmic objects at a period when the Universe was almost stationary. Indeed, let a light beam be emitted by a quasar during the epoch of suspended expansion. While this light propagates through the non-expanding Universe, it does not redden. Suppose further, that when the light beam passes by some other, closer quasar, the latter emits another light beam arriving at the telescope simultaneously with the first one. While the light beams are travelling in the almost stationary Universe, neither are shifted to the red. They both redden by the same amount later, when the expansion of the Universe is resumed. Hence, both a relatively close and therefore bright quasar, and a distant fainter one will have almost the same red shifts. In this way Petrosian, Salpeter and Szekeres tried to explain the absence of the m–z relation for distant quasars.

Even more convincing arguments in favour of the delayed expansion model have been presented by the Soviet astrophysicists I.S. Shkolvsky and N.S. Kardashev. The spectra of quasars revealed the following peculiarity. When the red shifts of emission lines in quasar spectra exceeded 1.95, the red shifts of absorption lines in the same spectra were, as a rule, equal to 1.95. I.S. Shklovsky and N.S. Kardashev explained the latter fact by assuming that the expansion of the Universe was delayed at an epoch from which the light arrives with a red shift of 1.95. Emission lines originate in quasars themselves and, consequently, their red shifts may be quite different. Absorption lines are formed not in quasars, but in the intervening galaxies, when the light travelling from a quasar to the observer happens to pass through some galaxy on its way. If the expansion of the Universe was suspended for a long time, the quasar light would most likely have crossed such an intervening galaxy just at the epoch of suspended expansion, which would result in the red shifts of absorption lines equal to 1.95. For this the repulsive acceleration must be represented by eq. (27), but with the numerical coefficient increased by a factor of 30. (This results from the fact that, according to the hypothesis of American and Soviet scientists, the force of gravity and the repulsive force must have cancelled each other out not now but far in the past, when the matter density in the Universe was about 30 times more than today.) Ten years have elapsed since this hypothesis was formulated. Much new observational data on quasars has been accumulated. On the one hand, it turned out that the quasars do not obey a strict $m-z$ relation because of a wide scatter in their intrinsic luminosities (see §8). If one isolates homogeneous groups of quasars with less scattered luminosities, the quasars within any such group show much more definite $m-z$ correlations. On the other hand, preliminary reports on the concentration of red shifts for the quasar absorption lines toward the preferred value $z = 1.95$ have not been confirmed. Again the Λ-term became unnecessary for the third time now.

But, as they say, once let out of a bottle, a genie is not easily forced back. The idea of a non-zero Λ-term had proved to be quite viable.

The only thing that is now clear is that the Λ-term, if not zero, is quite small. And, certainly, one will never be able to prove from the observations that it is exactly zero. Maybe a cosmic repulsion really exists after all?

This compels physicists to ponder over the possible nature of such a force.

One conceivable approach to elucidating the physical nature of the Λ-term is as follows. According to modern quantum theory, a vacuum is the lowest energy state of all physical fields. Even when there are no real particles at all, the continual creation and annihilation of so-called virtual particles takes place in a vacuum. Once created, the virtual particles cannot fly apart and become real particles that could be registered; instead, they, so to say, immediately annihilate on the spot. But the effects of the interaction of virtual particles can be detected, and they have been measured in laboratory experiments with a high degree of accuracy. An extremely complicated problem is that of the energy density of a vacuum. It turns out that the energy of the vacuum enters the formulae of the theory always in such a way that it is ultimately cancelled out, when the formulae are applied to real particle systems. The theory may be reformulated in such a manner that the average energy density of the vacuum becomes exactly zero. This approach, however, is justified only as long as the gravitational interaction of virtual particles is not taken into account.

During the third renascence period of the idea of the Λ-term (in the late 1960s) the Soviet physicist Ya.B. Zel'dovich put forward arguments that demonstrate in simple terms how a non-vanishing energy density of the vacuum could emerge. Virtual particles with a rest mass m (for simplicity we consider only one kind of particle) are being created and annihilated in the vacuum. The average density of proper mass (or of proper energy – the quantity differing from the mass density according to Einstein's formula $E = mc^2$ by the factor c^2 only) of virtual particles does not enter the final expressions and may be set equal to zero, as has been mentioned already. Quantum theory associates a characteristic length $l = \hbar/mc$ with any particle of mass m, where $\hbar$ is Planck's constant divided by 2π. The average separation of a newly born pair of virtual particles is about the characteristic length l. The energy of the gravitational interaction of such a pair can be estimated from the conventional formula:

$$E = Gm^2/l.$$

(See §5 of Chapter 5 concerning the sign of this quantity.)

It is this energy that can give rise to the non-vanishing energy density of the vacuum, or, correspondingly, to a non-vanishing mass density of the vacuum $\rho_{vac} = \varepsilon_{vac}/c^2$. To estimate the density of energy ε, we divide E by the volume l^3 occupied by one virtual particle

$$\varepsilon_{vac} = (Gm^2/l)/l^3 = Gm^6c^4/\hbar^4.$$

The last term in the above equation is obtained by the substitution of $\hbar/mc$ for l.

This is not all, however. The theory requires also that the 'vacuum fluid' exert some pressure, but, in contrast to pressure in the usual sense, this vacuum pressure must be negative. It would be even better, perhaps, to speak of the strain rather than pressure. The absolute magnitude of the vacuum pressure must be equal to that of the energy density, i.e. $p = -\varepsilon$; making use of the Einstein relationship, we can rewrite it as $p = -\rho c^2$, or $p/c^2 - \rho$. This property is very important and, as we shall see in a moment, may very well be the primary cause of the gravitational repulsion of the vacuum.

Recall that the mass of a uniform sphere is

$$M = \tfrac{4}{3}\pi R^3\rho.$$

But according to Einstein's theory this formula is valid so long as the pressure of matter is small compared with the quantity ρc^2. Otherwise the formula

$$M = \tfrac{4}{3}\pi R^3(\rho + 3p/c^2).$$

must be used. Under usual circumstances ρ greatly exceeds p/c^2, and the term $3p/c^2$ can be safely neglected. The mean matter density in the Sun for example is about 1.4 g cm^{-3}, while the mean pressure amounts to 2×10^{15} dyne cm^{-2}. Thus, the ratio $(3p/c^2)/\rho$ is about 10^{-6}, which is indeed quite small. This is not the case, however, for virtual particles in a vacuum, for which the pressure and the energy density of gravitational interaction are linked through the relation
$$\rho_{\text{vac}} = -p_{\text{vac}}/c^2.$$

It would be only natural to assume that the 'vacuum fluid' uniformly fills the whole of space. Therefore, to calculate the gravitational acceleration caused by such a 'fluid', we can use eq. (10) derived for the uniform distribution of ordinary matter, the only difference being that the density ρ must be replaced by the sum $(\rho + 3p/c^2)$. Recalling that $p_{\text{vac}} = -p_{\text{vac}}/c^2$, we obtain
$$a_{\text{vac}} = -\tfrac{4}{3}\pi G \, (\rho_{\text{vac}} + 3p_{\text{vac}}/c^2)R$$
$$= -\tfrac{4}{3}\pi G \, (\rho_{\text{vac}} - 3\rho_{\text{vac}}) \, R = \tfrac{4}{3}\pi G \, 2 \, \rho_{\text{vac}} \, R. \tag{30}$$

So, we have found that the gravity of a vacuum is not attractive, as for ordinary matter (the negative sign in eq. (10)), but, rather, repulsive (the positive sign of the last term in eq. (30)). Such a repulsion apparently stems from the fact that the vacuum pressure is negative and participates in the gravitational interaction, as Einstein's theory shows, on a par with the energy density.

As for the numerical value of the vacuum gravitational repulsion, the theory is too primitive as yet to allow for its more or less reliable calculation. Or, rather, the theory in the proper sense of the word simply does not exist, and only its most general outlines can be sketched as shown above. Only a very tentative estimation of the magnitude of the repulsion can be attempted. For this one has to express the density $\rho_{\text{vac}} = \varepsilon_{\text{vac}}/c^2$ in eq. (30) in terms of universal constants and substitute in the numerical values of these constants. The constants G, c and $\hbar$ are known to a high degree of accuracy. But we are quite uncertain about the value of the mass m of virtual particles (in fact the whole spectrum of possible kinds of virtual particles – rather than one species only – must be accounted for) which is to be substituted in eq. (30). At the same time, the ultimate result is very sensitive to the actual value of m, since it enters the formula in the sixth power. If one takes m to be equal to the proton mass, one obtains a numerical coefficient in eq. (27) for the repulsion acceleration some 10 million times greater than that inferred by I.S. Shklovsky and N.S. Kardashev from quasar observations. If m is chosen to be the mass of the electron, the ensuing numerical coefficient is 100 billion times less than that in eq. (27). Such a wide gap reflects the primitiveness of the theory. But the very fact that such crude estimates lead to values that are not all that far – according to astronomical standards – from conforming to the observations suggests that the first step has probably been taken in the right direction. It might be added that early attempts to evaluate the energy density and the gravity of a vacuum gave values exceeding those allowed by observations by a factor as large as 10^{46} – evidence that the physical ideas involved were wrong. Note that the fundamental theoretical principles do not forbid the energy density of a vacuum to be negative (rather than positive as we assumed in the text) and, consequently, its pressure to be positive. Then, as the net result, one gets instead of repulsion a gravitational attraction of the vacuum. The theory is at such an embryonic stage that it cannot assure even the sign of the effect! That is why we did not pay any attention to the sign of the vacuum energy density. The point of major significance, however, is that, whatever the sign of the vacuum energy density, its pressure must always be of the opposite sign and of exactly the same absolute magnitude. Otherwise, one would be able to point out a privileged reference frame at rest with respect to the vacuum – in contradiction to special relativity theory.

Thus, the question of whether a vacuum gravitates or not has yet to be answered. Even if the Λ-term is not zero, it is so small that it has no noticeable effect on the cosmological processes of today. The concepts of contemporary field theory have recently led to the belief that in the very first instants of the Universe expansion the properties of a vacuum were different from what they are now; ρ_{vac} was tremendous, the Λ-term was also big and strongly affected the beginning of the Universe expansion. The progress of the theory and new observations will show whether this hypothesis is valid.

12   The gravitational paradox

This chapter deals with the expanding Universe. It has been emphasized many times that the gravitational interaction implies non-stationary behaviour of the uniformly distributed matter of the Universe. All the reasoning has been conducted with the aid of

Newton's law of gravity only. But, as has been already mentioned in §2, Newton's law, when applied to the calculation of the gravitational field of all the matter in the infinite Universe, leads to a contradiction known as the gravitational paradox. The essence of this paradox is as follows. Let the Universe on average be uniformly filled with heavenly bodies, so that the mean matter density over large volumes of space is the same everywhere. Try to calculate from Newton's law the force of gravity, with which all the infinite matter of the Universe acts on a body (a galaxy, for example) placed at some arbitrary point in space. The result turns out to depend on the way in which the calculations are carried out. To demonstrate this, we employ the following chain of simple arguments. Assume first that the Universe is empty. Place a test body A at an arbitrary point of space. Surround this body with matter of uniform density making up a sphere of radius R, with the test body at its centre. There is no need to perform calculations to verify that, due to the symmetry conditions, the forces of gravitational attraction of material particles constituting the sphere are all balanced in its centre, and the resultant force acting on body A is exactly zero, i.e. no force at all is exerted on body A. Now begin to add to the sphere one after another spherical shells of the same density. We know already (see §2) that spherical shells of matter create no gravitational forces in their interiors. As a result, the addition of such layers alters nothing, i.e. the resultant force of gravity acting on A remains zero. Continuing the process of addition of new spherical layers up to infinity, we arrive at the infinite homogeneous Universe filled with matter of constant density, in which the resultant force of gravity acting on a test body A is zero.

But the reasoning might proceed the other way. Again, take a uniform sphere of radius R in an empty Universe. This time, however, place the test body not in the centre of the sphere with the same matter density as in the previous case, but, rather, on its surface. Now, according to Newton's law, the force of gravity acting on body A is

$$F = -GMm/R^2, \tag{31}$$

where M is the mass of the sphere and m is the mass of the test body A. Further, begin adding spherical shells of matter to the sphere under consideration. A spherical layer added to the sphere does not change the forces of gravity inside it. Therefore, the force of gravity acting on A will be F (given by eq. (31)) as before.

Continue adding spherical layers of matter with the same density. The force F remains unaltered. In the end we again arrive at the infinite Universe filled with homogeneous matter of the same density. But now the body A experiences a force F. Quite evidently, with an appropriate choice for the initial sphere one can get an arbitrary force F, on passing to the limit of the infinite homogeneous Universe. Such an ambiguity has been called the gravitational paradox.

So, Newton's theory does not provide the means to calculate unambiguously, without additional hypotheses, the forces of gravity in the infinite Universe. Only Einstein's theory makes it possible to carry out such calculations in a self-consistent way. At the end of this section we return to this question and discuss the reason why we were nonetheless able to employ Newtonian gravity to draw the conclusions of §2 and others.

Consider now why in cosmology, when exploring the enormous dimensions of the cosmos, we have to turn from Newton's theory to Einstein's.

For this let us first discuss why the necessity to modify Newton's law of gravity arises at all.

Newtonian theory assumes the instantaneous propagation of gravity, and for this reason alone cannot be reconciled with special relativity theory which implies that no physical interaction can propagate faster than light in a vacuum.

It is not difficult to derive the conditions limiting the applicability of Newtonian gravity theory. Since this theory is incompatible with special relativity, it fails always when the gravitational fields involved are strong enough to accelerate the particles travelling in them to velocities comparable with the speed c of light in a vacuum. The velocity acquired by a body falling freely from infinity (where it is assumed to have negligible velocity) to some point in space is of the order of the square root of the absolute magnitude of the gravitational potential ϕ at that point (at infinity ϕ is taken to be zero). Thus, Newtonian theory can be used only when

$$|\phi| \ll c^2. \tag{32}$$

Furthermore, this theory cannot be applied to calculate the motions of particles in weak fields (satisfying condition (32)) either, if those particles, while far away from gravitating masses, have already been moving with velocities comparable to c. In particular, Newtonian theory cannot be applied to calculate the path of light in gravitational fields. Though it predicts the effect of the bending of light rays by gravitational fields, the magnitude of the effect derived from Newtonian theory differs from that obtained in the framework of the relativistic theory of gravity. And it is the latter that has been confirmed by observation. Finally, Newtonian theory is not applicable to calculations of variable gravitational fields generated by moving bodies (binary stars, for example) at distances r exceeding $\lambda = c\tau$, where τ is a time scale characterizing the motions of the bodies (the orbital period of a binary star, for example). Newtonian theory in such cases demands an instantaneous adjustment of the gravitational field at a distance r to a snapshot arrangement of the moving bodies at the same moment of time, while the theory of special relativity forbids the change in the positions of moving bodies to be reflected in the change of the gravitational field at a distance r in a time interval shorter than that taken by light to cover the distance r.

The incorporation of the laws of special relativity into generalized gravitational theory was performed by Einstein in 1916. We discuss some conclusions of this theory in the next chapter.

Now let us come back to cosmology and demonstrate that here we face a situation where gravitational fields are strong enough to accelerate bodies to velocities comparable to the speed of light.

Consider once more a sphere of homogeneous material of arbitrary radius R and matter density ρ surrounded by a vacuum. If we now begin to increase the size of our sphere, keeping the density of its interior constant, the mass of the sphere will grow as the cube of its radius. On the other hand, the force of gravity at the surface of the sphere varies in inverse proportion to the square of its radius. The product of these two factors results in the force of gravity increasing linearly with the radius of the growing sphere. Clearly, continuing to infinity we sooner or later arrive at such a strong gravitational field on the surface of the sphere that Newtonian theory is invalid.

The above argument can be expounded in greater detail to provide an estimate of the distance in our real Universe, at which Newtonian theory can no longer be used. The velocity which a body acquires when falling freely from a large distance, where it was initially at rest, onto the surface of a gravitating sphere is, apparently, exactly the same as that needed by the body to escape from the surface of this sphere to infinity, namely the escape velocity. Let us again write down the formula for the escape velocity:

$$v = (2\ GM/R)^{\frac{1}{2}}. \tag{33}$$

If the mass of the sphere is large enough compared to its radius, the free fall velocity may reach the speed of light. Substituting c in place of v in eq. (33), we find that the value of the critical radius of a sphere of mass M is given by

$$R_{\text{crit}} = 2\ GM/c^2. \tag{34}$$

In general relativity theory this radius is called the gravitational radius.

Now expressing the mass of the sphere in terms of its matter density and volume, $M = \frac{4}{3}\pi R^3\ \rho$, we get

$$R_{\text{crit}} = (3\ c^2/8\pi G\rho)^{\frac{1}{2}}. \tag{35}$$

To infer a numerical value of R_{crit}, one has to insert in the above expression the numerical value of ρ.

The mean matter density ρ in the Universe is known from observations, though not to a particularly high degree of accuracy (see §9). For an order-of-magnitude estimate we can adopt the value $\rho = 10^{-30}$ g cm^{-3}. Now, with all the quantities on the right-hand side of eq. (35) known, we can find the critical distance:

$$R_{\text{crit}} \approx 4 \times 10^{28}\ \text{cm} \approx 10^{10}\ \text{pc}. \tag{36}$$

Thus, when distances in excess of about ten billion parsecs are considered in astronomy, one must use the general theory of relativity.

An attentive reader will almost certainly wonder why, all the same, we can use Newtonian theory for smaller distances. As was demonstrated at the beginning of the section, Newtonian theory – he (or she) will argue – leads to the gravitational paradox (to an ambiguous value for the force of gravity in the Universe) and, hence, cannot be trusted on any scale at all? Strictly speaking, the reader would be quite right. Whatever the scale, Newton's theory of gravity by itself would never make it possible to obtain correct results in an at least somewhat convincing way. But if the reader rereads §2, he will notice the following statement substantiating the applicability of Newtonian theory on small scales in the Universe: Einstein's relativistic theory of gravity (which is concisely described in the following chapter and which can be applied to the whole of the Universe without any paradoxes) ascertains that a spherically symmetric distribution of matter (regardless of whether it is finite or infinite) creates no gravitational field inside a spherical cavity (see Fig. 17). This conclusion of Einstein's theory is, in contrast to that of Newton's theory, quite rigorous and obtained in an unambiguous way. Using this, one can consider a sphere filling such a cavity with matter of the same density as that of the surrounding material. If the sphere is small enough and its own gravitational field is weak, one can use Newtonian gravitational theory. Such reasoning is legitimate and rigorous, and we have successfully made use of it, avoiding the gravitational paradox.

To all this one need only add that the quantities that can in principle be measured in cosmology are the relative velocities of galaxies and their relative accelerations. The gravitational force F takes on a specific value only after one has fixed the position of the centre of the sphere, or, as is said in relativity theory, after one has chosen a reference frame. But the position of the sphere's centre, or the reference frame, can be chosen arbitrarily! That is how the gravitational force F – as well as other concepts of Newtonian theory including length and time – becomes a relative quantity in relativity theory.

The above discussion can be summarized as follows. Fo. compara-tively small regions in space, less than about ten billion parsecs in extent, one can safely use Newtonian theory to tackle cosmological problems. In particular, this theory elucidates the evolutionary behaviour of the Universe. On scales comparable to and greater than the critical length – and even more so in the analysis of the structure

Fig. 17. An infinite spherically symmetric distribution of matter creates no gravitational field inside a spherical cavity.

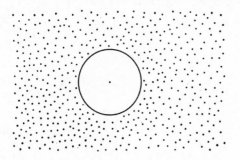

of the Universe as a whole – one needs the relativistic theory of gravity. Einstein wrote: 'From my point of view one cannot arrive, by way of theory, at any at least somewhat reliable results in the field of cosmology, if one makes no use of the principle of general relativity'.[†]

In the next chapter we discuss briefly the relativistic theory of gravity and relativistic cosmology.

[†] See *Albert Einstein: Philosopher–Scientist*, p. 684, edited by P.A. Schilpp, Library of Living Philosophers, Evanston, Ill.

2

Relativistic cosmology

1 The principal idea of Einstein's gravitational theory

We begin this chapter with a brief description of key ideas and basic conclusions of Einstein's general theory of relativity (relativistic theory of gravity), which the Soviet physicists L.D. Landau and E.M. Lifshitz described as 'the most beautiful of all existing physical theories'. Of course, a detailed account of the theory itself is not intended here. Without any reference to the complicated mathematical apparatus of the theory, we present the most important results which are indispensable to the comprehension of today's cosmological problems. (For the foundations of Einstein's gravity theory see §5 of Chapter 5.)

The most important property of gravity, discovered by Galileo and Newton and laid as the foundation of the new theory by Einstein, is that it acts in the same way on all bodies, imparting an equal acceleration regardless of mass, chemical composition or any other characteristics of different objects. This property had been established experimentally by Galileo and can be formulated as the principle of the gravitational mass m_w (which determines the interaction of a body with gravitational fields and enters the universal law of gravity; see eq. (31) of Chapter 1) of any body being exactly equal to its inertial mass m_i (which determines the resistance of the body to external forces and enters the second law of Newtonian mechanics). To illustrate this principle, we state the equation of motion (the second law of Newtonian mechanics) of a body in the gravitational field of a spherical mass M_w:

$$m_i a = F = -m_w \, GM_w/r^2, \tag{1}$$

where a is the acceleration received by the body due to the attraction of the gravitating sphere. If m_i is equal to m_w, they both drop out of eq. (1); as a result, the acceleration a does not depend on the mass of a test body and is determined by the strength of the gravitational field only (given by the term GM_w/r^2), as first discovered by Galileo. Modern laboratory experiments confirm Galileo's law to a very high degree of accuracy. For example, the Soviet physicist V.B. Braginsky has demonstrated that m_i and m_w agree to better than one part in 10^{12}!

Since our school days we have been used to the fact that the mass m_i in the second law of mechanics is the same as the mass m_w in the law of gravity. But this is merely from habit! If, instead of the gravitational interaction, we discussed the motion of a test electric charge in the electrostatic field of some other charged object, the left-hand side of eq. (1) would contain m_i as before, while the masses on the right-hand side would be replaced by electric charges. It is the charges that determine the force. Thus, m_w may be called the gravitational charge! And the remarkable fact, which must be explained, is that this 'charge' turns out to be equal to m_i. More precisely, one must speak not of the equality between m_i and m_w, but rather of the proportionality relation between them. Indeed, if m_i and m_w were not equal but only proportional to each other in eq. (1), the free fall acceleration would all the same be independent of mass. The exact equality of m_i to m_w could then be arranged by a suitable choice of units in which to measure m_i and m_w. But these are subtleties of no particular significance here.

Thus, bodies with different masses or of a different physical nature move in exactly the same way in a fixed gravitational field, provided that their initial velocities are the same. This fact is a manifestation of a deep analogy which exists between the motions of bodies in gravitational fields and the motions in field-free regions as viewed by accelerated observers. When there is no gravity, the bodies with different masses all move by inertia with constant speeds along straight trajectories. But if the observer watches these same bodies from the cabin of a uniformly accelerated spaceship, he will see all of them as moving with one and the same constant acceleration, opposite in direction to the acceleration of the spaceship. The motions of the bodies will be indistinguishable from free fall in a uniform gravitational field. Therefore, an accelerated reference frame (associated with the spaceship in our case) is equivalent to a gravitational field. This statement embodies Einstein's equivalence principle. According to this principle one can perform the reverse procedure as well (as contrasted to the imitation of gravity with an accelerated reference frame) – namely, to 'get rid' of the true gravitational field at a given point by introducing a reference frame moving with the free-fall acceleration. This is clearly demonstrated by the well-known fact that things have no weight – do not 'feel' any gravitational field – in the cabin of a spaceship orbiting the Earth with its engines turned off. Einstein suggested that not only mechanical motions but all physical processes in a true gravitational field are governed by exactly the same laws as in an accelerated reference frame without any gravitational field at all. This principle has been called the 'strong equivalence principle' as contrasted with the 'weak equivalence principle' concerning motions in space only.

The above reference frame (a spaceship powered by a rocket engine) moving with constant acceleration in the absence of gravitational fields imitates a uniform gravitational field, which is of the same magnitude and direction everywhere. Gravitational fields generated by individual bodies are not like that. To imitate the spherically symmetric gravitational field of the Earth, for example, one has to introduce accelerated reference frames with accelerations directed differently at different points. Having established a communication system, observers in different frames will find out that they are accelerating with respect to one another, and will ascertain in this way the absence of a true gravitational field. Thus, a true gravitational field is not equivalent to a simple accelerated reference frame in normal space, or, more precisely, in the spacetime of the special theory of relativity. Einstein showed, however, that if, insisting on the equivalence principle, one requires the true gravitational fields to be equivalent at every point to local accelerated reference frames, the global spacetime becomes curved, i.e. non-Euclidean. This means that the geometry of three-dimensional space becomes non-Euclidean; the sum of the angles in a triangle is no longer π, the ratio of the circumference to the radius in a circle is no longer 2π, and so on; time at different points flows at different rates. In short, Einstein's gravitational theory claims that the true gravitational field is nothing more than a manifestation of the curvature (the departure of the geometry from the Euclidean one) of four-dimensional spacetime. It is not incidental that we speak of four-dimensional spacetime, since special relativity theory (which does not treat gravity) had already united space and time into one physical entity.

This theory has been described in many popular books and we shall not dwell on it here.

It should be emphasized that Einstein's theory of gravity could be created only after N.I. Lobachevski, J. Bolyai, B. Riemann, K.F. Gauss and others had constructed non-Euclidean geometrical models.

Without gravity, in the spacetime of special relativity, bodies move by inertia along straight lines, or, as mathematicians put it, along extremal lines (geodesics). In flat spacetime any extremal line is a straight line.

The idea, laid down by Einstein as the foundation of the relativistic theory of gravity and linked closely to the equivalence principle, is that in the presence of gravitational fields all material bodies move along extremal lines (geodesics) of the spacetime which is, however, no longer flat and in which the geodesics are no longer straight. Of course, the curvature of space – and more so the curvature of four-dimensional spacetime – cannot be visualized in a simple pictorial manner. But physicists have long since been used to working with completely non-pictorial concepts (think of quantum mechanics, for example). The curvature (the non-Euclidean character) means only that the geometry of the manifold is defined by axioms other than those of Euclid. The properties of the extremal lines in such manifolds differ from the properties of straight lines. As an example of extremal lines on a two-dimensional curved surface one can point out the arcs of great circles on the surface of a sphere.

Masses generating gravitational fields cause spacetime to be curved. Bodies moving in such a curved spacetime travel along the same geodesics, irrespective of the masses or chemical compositions. The observer interprets these motions as the motions along curved trajectories in three-dimensional space with variable speeds. It is inherent in the very heart of Einstein's theory that the bending of trajectories (the rate of change of velocity) is entirely the property of the spacetime – the property of the geodesics in this spacetime; hence, all bodies must be accelerated in the same way; and hence, to stipulate the equality of accelerations for different bodies in eq. (1), the gravitational mass m_w must be equal to the inertial mass m_i.

Thus, the gravity field is nothing more than a departure of the properties of spacetime from the properties of a flat manifold.

Einstein's gravity equations link the quantities describing the curvature of the spacetime (i.e. the strength of the gravitational field) with the quantities characterizing the generators of the curvature (i.e. the sources of the gravitational field). According to Einstein's theory, gravitational fields are generated not only by mass (as in Newton's theory), but also by the kinetic energy of matter in motion, by its pressure and by its other characteristics in more complicated surroundings. All kinds of matter experience the action of gravity (which is, in fact, the curvature of the spacetime), and all kinds of matter participate in the generation of the gravitational field. The gravitational field is, for example, affected by the presence of electromagnetic fields and by other physical fields as well.

As mentioned above, we do not intend to write out here Einstein's equations in their most general form. In any case, we would be unable to explain how they are solved and used in applications, without employing the complicated mathematics of general relativity theory. What has been said is quite enough to understand further discussions. Occasionally, when needed, we shall formulate results derived from the equations of this theory.

There should be no doubt that in the limit of weak gravitational fields Einstein's equations reduce to Newton's universal law of gravity, while the spacetime departs only slightly from the flat (Euclidean) one.

2   The geometric properties of space in the Universe

Let us return now to cosmology and consider regions of space with dimensions comparable to R_{crit} (as defined in § 12 of Chapter 1) and even greater, which can be studied in the framework of relativistic theory only.

We shall discuss large volumes of space in which matter can be regarded as being uniformly distributed (see Chapter 1) and all the properties are independent of direction. As was mentioned in the Introduction, a cosmological model with these assumptions, based on Einstein's equations, was first constructed by A.A. Friedmann.

Although our interest here will be focused on the geometry and curvature of three-dimensional space, one should not forget that in general relativity theory not only is three-dimensional space curved, but so is the whole of four-dimensional spacetime as well.

We point out once more, that all these complications as compared to classical notions of Newtonian mechanics are absolutely necessary. Newtonian mechanics dealt with the absolute space in which material bodies moved. The special theory of relativity rejected the idea of absolute space and absolute motion and demonstrated that, in order to define the motion, one has to introduce a reference frame. Only after the reference frame is specified, does the description of motion become meaningful. The general theory of relativity demonstrated that spacetime is curved in proportion to the gravitational field (in fact, gravity is a manifestation of the spacetime curvature). If the gravitating substance moves, the curvature changes with time. As a result, any reference frame changes – deforms – with time. So, there can be no rigid structure, no rigid frame of permanent spatial coordinates – they will all inevitably warp in variable gravitational fields.

It goes without saying, that in weak gravitational fields all the deformations of a coordinate system due to the changes in the spacetime curvature can be made negligibly small. This is far from being the case in cosmology, however, where big masses are involved; the spacetime curvature varies substantially and a rigid coordinate system is impossible. This clarifies the statement made in §4 of Chapter 1 that the concept of recessional velocity loses its habitual unambiguous meaning for very distant galaxies. One cannot, indeed even in principle, extend imaginary rigid rods to remote galaxies and measure their velocities with respect to these rods. Special relativity theory established that no absolute space and no absolute velocity exist. But according to this theory, once the observer is fixed, one can associate with him a rigid coordinate system and measure the velocities of bodies with respect to this observer at arbitrarily large distances. General relativity theory goes further and asserts that in gravitational fields, even once the observer is fixed, one cannot nevertheless attach to him an extended rigid coordinate system and determine unambiguously the velocities of remote galaxies (for such distances inevitable deformations of the coordinate system are quite large). But the inability of Einstein's theory to define unambiguously the recession velocities of galaxies does not preclude it from ensuring the exact and unambiguous calculations of the red shifts of spectral lines in the spectra of remote galaxies, their apparent brightnesses and other observable quantities.

After this brief digression, we come back to cosmology. In Friedmann's homogeneous isotropic cosmological model the reference frame can be chosen quite naturally. Imagine a coordinate system which is like a cobweb attached to the galaxies and stretches out as the entire system of galaxies (or, rather, clusters of galaxies) expands. The galaxies are at rest with respect to this coordinate cobweb. (Apart from their common motion due to the cosmological expansion, the galaxies have random velocities which usually amount to hundreds (sometimes thousands) of kilometres per second. We do not pay any attention to these random velocities, considering only the average motion of the 'galaxy gas'.) Such a reference frame is called a 'comoving system'. We adopt this system of coordinates and come to the study of the geometric properties of 'comoving space' in this reference frame.

What is the curvature of space? It is a distortion of its geometric properties. One can readily imagine a curved two-dimensional

Fig. 18. Two-dimensional curved surfaces. Above: a sphere – a surface of positive curvature. The sum of angles in a triangle upon the sphere is greater than π. Below: a hyperboloid of one sheet – a surface of negative curvature. The sum of angles in a triangle upon the hyperboloid is less than π.

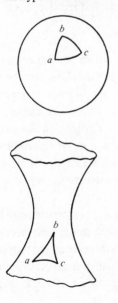

surface (as illustrated by the surface of a sphere or by the surface of a hyperboloid, see Fig. 18). On such a surface the geometry is different from that of the plane. But this analogy will be of hardly any help when trying to visualize curved three-dimensional space. We live in this three-dimensional space and cannot leap out of it because there is nothing beyond it; therefore, one cannot ask the question: 'into what is space bent?' (although mathematicians may consider curved three-dimensional space as being embedded in abstract spaces of higher dimensions).

The essence of spatial curvature is in its geometrical properties being different from those of flat space, where Euclidean geometry is valid.

In curved space planes are replaced by geodesic surfaces, the straight lines by the geodesic lines. As has been already mentioned, the sum of angles in a triangle made by geodesics in curved space is not π (or $180°$) and depends on the area of a triangle, the circumference of a circle is not $2\pi r$, and so on. The curvature of a two-dimensional surface at a given point is characterized by one number only. This number is defined as follows. One measures the sum Σ of angles of a small triangle surrounding a given point. On a curved surface this sum will be either more or less than π. It can be proved that the difference between Σ and π is directly proportional to the area S of the triangle:

$$\Sigma - \pi = CS \tag{2}$$

The proportionality coefficient C is called the curvature. The curvature C may be either positive or negative. The square root of the quantity $1/C$ is called the radius of curvature. When C is negative, the radius of curvature is imaginary. For example, the curvature of a spherical surface (which is the same at all points on the surface) is positive and equal to the inverse of the square of the radius of the sphere; the radius of curvature is the radius of the sphere. As an example of a surface with negative curvature, we can use a hyperboloid of one sheet; see Fig. 18.

The curvature of three-dimensional space is a much more complicated notion. But we are actually interested in the simplest case of a homogeneous and isotropic Universe. In this case spatial curvature is also characterized by one number only, which is called the curvature and defined in exactly the same manner as for a surface. Equations of general relativity theory yield that the curvature of space is negative when the matter density in the Universe is below its critical value (see Chapter 1) $\rho_{\text{crit}} = 10^{-29}$ g cm^{-3}, and positive in the opposite case.

The sign and magnitude of the curvature in a homogeneous Universe at any fixed time are the same everywhere in space. The value of the radius of curvature depends on the values of the Hubble constant H and matter density ρ, namely

$$A = (c/H) \, [\rho_{\text{crit}}/(\rho - \rho_{\text{crit}})]^{\frac{1}{2}}. \tag{3}$$

Here c is the speed of light. If, for example, the density ρ is twice the critical one, the radius of curvature is

$$A = 4 \times 10^9 \text{ pc}. \tag{4}$$

If the matter density ρ is less than the critical value, the radius of curvature becomes imaginary and the curvature itself (being the inverse of the square of the radius of curvature) is negative. In this latter case the bending of space is characterized by the absolute magnitude of the imaginary radius of curvature. With the matter density and the value of Hubble constant varying with time, the curvature radius is also variable, but the sign of the curvature remains the same during the whole evolution of the Universe. The radius of curvature changes with time in the same manner as the separation of any pair of galaxies does. For this reason the plots of Fig. 9 depict the variation with time of the quantity A as well. In the case $\rho < \rho_{\text{crit}}$ the absolute magnitude of A starts from zero at the outset of the expansion and then grows without limit. In the case $\rho > \rho_{\text{crit}}$ the radius A grows, attains its maximum value at some time, and drops to zero afterwards.

3   Closed and open universes

Let us look at the specific conclusions that can be drawn from the fact that space may be curved. If the spatial curvature of the Universe is negative (i.e. if the actual matter density is less than the critical one, $\rho < \rho_{\text{crit}}$) or zero, space extends to infinity in all directions and contains an infinite number of celestial bodies, elementary particles and galaxies. Such a Universe is called an open Universe. Qualitatively, the structure of this Universe is similar to that of simple Euclidean space.

If, however, $\rho > \rho_{\text{crit}}$ and the spatial curvature is positive, three-dimensional space turns out to be closed and finite (though unbounded). Again, we wish to remind the reader that all our pictorial images are based on everyday experience and are closely related to three-dimensional Euclidean space. Therefore, it is impossible to visualize a closed Universe in a pictorial way; one can only explore its properties mathematically, compare the results of calculations with experimental and observational data, and illustrate them with analogies and models – but, of course, no model and no analogy can be perfectly adequate. So, what are the properties of the closed Universe? Pick some point (our galaxy, for example) as the origin of a coordinate system. Isolate an imaginary sphere around this point, i.e. consider a set of galaxies residing at the same distance from the galaxy chosen as the origin of our coordinate system. On the surface of this sphere, one can measure its total area, then draw an equator and measure its total length. While doing this, one ought not to forget that the Universe is non-stationary. The Universe is expanding, and the surface of the sphere fixed on the same galaxies varies in size. The length of the equator and the surface area of the sphere encompassing a fixed number of galaxies depend on the time at which they are measured. We want all the measurements to be carried out simultaneously, and in order to specify the notion of simultaneity we introduce the concept of proper time. This time is reckoned from the moment of infinite density by clocks moving alongside the particles. We consider here only the motion of the particles that corresponds to the expansion of the homogeneous Universe as a whole, i.e we neglect possible deviations of actual galactic velocities from the values that are assigned to every point by the law of expansion. It must be agreed now that all the quantities are being measured simultaneously for all galaxies, i.e. at one and the same moment of the proper time.

We arrive, then, at the following property of the closed Universe (where $\rho > \rho_{crit}$): as more and more distant spheres are considered at the same time, the length of the equator and the surface area first increase as in familiar flat space, then reach certain maximum values and ultimately decrease back to zero.

Thus, at large enough distances a sphere, being more distant from the origin of coordinates, containing more matter and having a greater volume, possesses at the same time a shorter equator and a smaller surface area. This appears to be very strange; quite unlike anything in Euclidean geometry, but it is a direct consequence of the curvature of space.

To illustrate the above description, let us turn again to analogies with curved two-dimensional surfaces. A two-dimensional analogue of closed three-dimensional space is the surface of a usual three-dimensional ball. Take its north pole to be the reference point. Analogues of the spheres in three-dimensional space are circles on the surface of the ball drawn around the north pole, i.e. geographic parallels. First, the length of the circumference of a circle increases as we move away from the north pole, then it reaches its maximum value at the equator and starts to decrease as we move further away, while the surface area of the ball encompassed by the same circumference continues to grow monotonously. Finally, when the circumference approaches the south pole, the surface area encompassed tends to $4\pi r^2$, while the length of the circumference itself tends to zero.

Notice that any parallel divides the ball's surface into two parts, both of which are finite. We always assume that the surface area encompassed by a parallel lies to the north of it and includes the north pole. Under such a definition the surface area encompassed by a parallel tends to $4\pi r^2$ as the latter tightens around the south pole.

Thus, we see that two-dimensional space of constant positive curvature turns out to be finite in surface area, closed onto itself, and has no boundary at all. A fly running along the surface of the sphere will never meet an edge or any other obstacle. No straight lines exist on the surface of the sphere. The shortest line (a geodesic) connecting two points on the spherical surface is an arc of a large circle. If one starts from the north pole and travels along a geodesic (a large circle), one eventually encircles the whole sphere and comes back to the initial point at the north pole from the other side.

Let us return to closed three-dimensional space which corresponds to a homogeneous Universe with $\rho > \rho_{\mathrm{crit}}$. Here we also define the interior of a sphere around the origin of the coordinate system as that part of the whole of space which contains the origin of coordinates and is bounded by the surface of the sphere. As mentioned above, when more and more remote spherical surfaces are considered, they first increase in size, then halt and start to decrease, ultimately shrinking to a single point. And this means that the three-dimensional space under consideration is closed. In a closed homogeneous

Universe a geodesic starting from the origin of a coordinate system – like a great circle on the surface of a sphere starting from the north pole – goes all the way round the finite Universe and returns to the point of origin, i.e. the 'straight lines' (geodesics) are in fact closed. The total length of such closed geodesics, expressed in terms of the radius of curvature of space A, is $2\pi A$.

We wish to emphasize once more that the closed Universe is, nevertheless, unbounded. There is no boundary in a finite space that one would be unable to cross. Nothing exists beyond this closed space. The volume of the closed Universe is finite. The Universe contains a finite number of heavenly bodies, galaxies, particles. The total volume of the closed Universe can be expressed in terms of its radius of curvature A:

$$V_{\text{Universe}} = 2\pi^2 A^3. \tag{5}$$

The value of the radius of curvature is determined by the values of ρ and H, as given by eq. (3). Recalling that A changes with time, we conclude that the volume of the Universe V is also variable. The volume is equal to zero at the beginning of the expansion; then it increases with time, reaches a maximum value together with A and shrinks back to zero again. Thus, it is of major significance to be able to determine whether the matter density exceeds the critical value not only because this will shed light on the future evolution of the Universe, but also because it will tell us about its spatial extension.

To conclude this section, we add the following remark. Exploring the case of the closed Universe, we inevitably arrive at the concept of a space with topological properties completely different from those of normal Euclidean space. It should be emphasized, however, that the topological properties of space (i.e. the most general properties of space as a whole) are not uniquely determined by its curvature alone, or, in other words, by its internal geometry alone. To illustrate this, consider another two-dimensional analogue. Take an infinite plane with Euclidean geometrical properties. One can bend it any way possible (but without either pleating or tearing it) and the properties of geometrical figures on it do not change, i.e. the geometry remains Euclidean – the angles of a triangle add up to π, the curvature $C = 0$. Now, cut out of this plane an infinite band and roll it up into an infinitely long cylinder (Fig. 19). Both the plane and the cylinder have zero curvature, $C = 0$, both have no boundaries (the fly can run along the cylinder meeting no edges, just the same). But their topological properties are quite different! For example, some geodesics on the cylinder are closed (see Fig. 19), others are infinite, etc.

Other, more complicated topological constructions are possible. Thus, the above simple two-dimensional example demonstrates that the curvature alone does not describe uniquely all the properties of a surface. In addition, the topology must be specified too. The same is valid for three-dimensional space (and for spaces of higher dimensions) as well. The general theory of relativity defines the geometry of space (or rather that of the four-dimensional spacetime) and says nothing in general as to its topology.

The discussion of this important issue is continued in § 6 of Chapter 5.

Fig. 19. A plane extending to infinity and an infinite cylinder – two boundless surfaces with the same curvature $C = 0$ but with different topological properties.

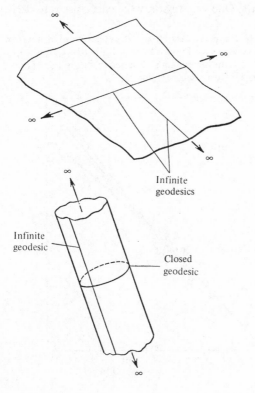

Here we just emphasize once more that for the simplest and most natural topologies the Universe is spatially infinite in the case $\rho \leqslant \rho_{crit}$, and the Universe is necessarily closed and finite in the case $\rho > \rho_{crit}$.

4   The mean matter density of the Universe and observational tests for the curvature of space

The curvature of the three-dimensional space can be calculated directly if one knows the Hubble constant and the mean matter density. Unfortunately, the matter density, as we have seen already, is so poorly known that we cannot even determine the sign of the curvature – whether it is positive or negative.

Perhaps one could reverse the problem? Is it possible first to measure the curvature of space, and then, with curvature known, to calculate the matter density ρ? In so doing one would be able to determine ρ with all forms of matter automatically accounted for – barely detectable kinds of matter equally as well as the conspicuous ones. One way to study the curvature is to derive from

Fig. 20 (*a*) The *m–z* relation with theoretical curves drawn for different values of the matter density. 1: $\rho = 10\,\rho_{crit}$, 2: $\rho = 5\rho_{crit}$, 3: $\rho = 2\rho_{crit}$, 4: the dependence obtained for the model of the stationary Universe, see 5 of Chapter 5.

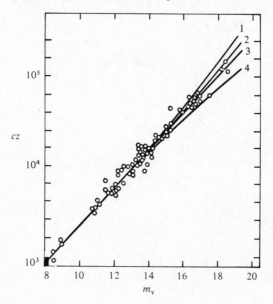

observations the m–z relation for very distant galaxies (or quasars). Calculations using the formulae of relativistic cosmology yield that the relation m–z for very distant objects deviates from the straight line discussed extensively in Chapter 1 and derived within the framework of classical mechanics and electrodynamics. The magnitude of this deviation depends on the curvature of space. Fig. 20a shows m–z curves corresponding to different values of the spatial curvature and, consequently, to different values of the matter density ρ. From this figure one sees that the existing observational data do not enable us to decide as yet whether ρ exceeds ρ_{crit} or not. Another way to determine the curvature of space is to count the number of galaxies (or radio sources) with apparent stellar magnitudes below some fixed value.

Fig. 20 (*b*). A scheme for measuring the angular sizes of objects.
  I. The angular size of an object in empty space. The farther away the object, the smaller the angle under which one sees it.
  II. Angular sizes of objects in the Universe filled with a translucent matter. Light rays propagating from an object to the observer are deflected by the gravitational attraction of matter inside the light cone. A remote enough object will subtend a larger angle than a closer one.

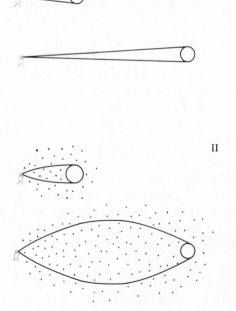

The dependence of this number on the fixed value of the stellar magnitude is different in spaces with different curvatures. In this method, however, the intrinsic luminosity of galaxies has to be known, or, at least, must be the same for all objects. Many technical difficulties hamper this line of study. But apart from these technical difficulties there is one of principle too. The light that we receive from distant objects was emitted eons ago, i.e. this light was emitted by objects at much earlier stages of their evolution than that of nearby objects. For this reason one has to account for the evolution of the galaxies – the changes in their luminosities, spectra, etc. – when trying to deduce the value of ρ from plots similar to that of Fig. 20a. How galaxies evolve remains practically unknown as yet, and no definite conclusion as to the value of ρ can be obtained in this way at present.

Big hopes in cosmology have been associated with radio astronomy. Radio telescopes are extremely sensitive instruments. But radio emissions of galaxies proved to be subject to evolutionary changes even more than their optical radiations, and radio astronomy gave no solution to the problem of spatial curvature. The issue still remains on the agenda of the natural sciences.

Here we wish to point out an interesting phenomenon that must occur in the homogeneous Universe. Suppose we observe objects of the same linear size (similar galaxies, for example) at different distances in the usual flat and empty space; the farther away the object, the smaller the angle it subtends (Fig. 20b). The angular sizes of objects decrease as the inverse of their distances.

A different picture is found in space filled uniformly with transparent matter – with a transparent gas or overlapping galactic coronae, for example. From general relativity theory one knows that light rays passing gravitating bodies are deflected, attracted by those bodies. Of the same origin is the well-known deflection of $1''.75$ experienced by the star light passing near the surface of the Sun. For this same reason the gas inside a cone of light rays from a distant galaxy will bend the rays inwards as shown schematically in Fig. 20b. The farther away the galaxy, the greater the mass that accumulates within the cone, the greater is the deflection of light rays. As a result, beyond a certain distance, more remote objects have greater, rather than smaller, angular sizes than those nearer to us! Of course, this effect occurs only if the cone of light rays contains gas (or some other matter). In the opposite case, when all the matter has condensed into galaxies and the intergalactic space is empty, no matter gets into the light-ray cone and the above effect vanishes, despite the high average density of the galactic matter. This was pointed out by Ya.B. Zel'dovich.

5   The observable horizon in the Universe

Of principal importance to observations in cosmology and to physical processes that occurred in the remote past in the course of expansion of the Universe is the existence of the so called observable horizon. The existence of this horizon is due to the expansion of the Universe. The farther away a galaxy is, the more time the light takes to reach the Earth. The light that reaches the observer today left the galaxy long ago. The Universe began to expand some 10–20 billion years ago. The light emitted by some source even at the moment of the start of the expansion is able to cover only a finite distance in the Universe – namely, the distance of some 10–20 billion light years, or about $(3–6) \times 10^9$ pc. Points in the Universe at this distance from us make up the observable horizon. The parts of the Universe beyond the observable horizon cannot even in principle be observed today. We cannot see more distant galaxies; no matter how good our telescopes are, light from galaxies beyond the horizon simply has not had enough time to reach us. The red shift of an object tends to infinity as the object approaches the horizon. The objects on the horizon itself have infinite red shifts. Thus, we can see a finite number of galaxies and stars in the Universe. In this way one more paradox of the pre-Einsteinian cosmology was resolved – the photometric paradox. It goes as follows. In the infinite Universe, filled with stars, any line of sight must meet eventually the shining surface of some star. As a result, the whole sky would be as bright as the surface of the Sun and other stars. In reality, however, we see a finite number of stars rather thinly scattered in space, and, consequently, the night sky appears dark. In addition, the stars have limited lifetimes.

The observable horizon makes the difference between the closed and open Universes less dramatic. In both cases we can see only a bounded portion of the Universe with a radius of some 10–20 billion light years. Even in the closed Universe the light has not had enough time to complete its round-the-Universe journey yet, and, of course, one cannot see the light emitted by our galaxy that has encircled the whole of the Universe. In the closed Universe one cannot see the 'back of one's head'. Even for the whole of the expansion period, from the singular state up to the turning point when the expansion reverses to a contraction, the light has only enough time to cover half of the closed space, and only during the contraction phase will it finish its round-the-Universe trip.

Note that, in this respect, the Universe with Λ-forces of repulsion (when the expansion could have been suspended for some time – see § 11 of Chapter 1) may significantly differ from the Universe without Λ-forces. The cosmological model with the Λ-term described above must necessarily have space of positive curvature and be closed. The forces of gravity can slow down the expansion of the Universe and bring it to a prolonged halt if only the matter density ρ is high and, consequently, the curvature of space is positive. In the model with a delay the Universe is almost static for a long period of time. This time may turn out to be quite sufficient for the light to go all the way round the closed space and come back to the initial point. If the delay in the expansion is long enough, the light can round the Universe many times. In this case one would be able to see a number of images of one and the same object – a galaxy, for example. Perhaps we actually see such 'ghost' images on the sky and take them erroneously to be different galaxies? We already argued at the end of Chapter 1 that it is most unlikely for such a delay in the expansion to have occurred in the real Universe. But in principle such a model is quite conceivable!

Any observer in the Universe has his own observable horizon. There are no preferred points in the Universe. The horizon of every observer expands with time; light has time to come to an observer from more and more distant parts of the Universe. For every hundred years the radius of the horizon increases by one part in a hundred million.

We make one further remark. Near the horizon one should see matter from the remote past, when its density greatly exceeded the present value. No individual objects existed at that epoch, and the matter was opaque to light. We shall return to this issue in the next chapter.

Finally, note the following. The very concept of a theoretically observable horizon could be introduced because the cosmological models defined a moment in the past when the expansion started, when the matter density was $\rho = \infty$, and the finite time interval that had elapsed since that moment was not long enough for the light to reach the observer.

If the Universe near the singularity $\rho = \infty$ did not expand as Friedmann's models predict (see Chapters 3 and 5 on this subject), the laws of light propagation could have been quite different at those times; the light – under certain conditions – could have travelled vast distances near the singularity and there might then be no theoretical horizon at all. Finally, if a contraction stage preceded the singularity $\rho = \infty$ (see Chapter 5 on this), no theoretical horizon would exist because the light emitted prior to the moment of $\rho = \infty$ would traverse a greater distance than the light emitted exactly at $\rho = \infty$. The real light would of course inevitably get absorbed at times of high density ρ, but we discuss the theoretical horizon, for hypothetical particles passing freely through the dense matter. For such particles no horizon would exist. (We assume, of course, that these 'superpenetrating' hypothetical particles travel with the speed of light.)

As we shall see later, all these possible versions which predict no horizon are rather unlikely and probably have nothing to do with reality. And without doubt, any variant of the cosmological model which pertains to reality must have an 'observable horizon in practice' arising from the fact that, peering far out into space, one looks far back in time and cannot see anything beyond the early dense stage of the expansion at which no real particle could propagate freely. Any signal from the remotest parts of space – even if there were time enough for it to reach the observer in principle (suppose the signal had been sent at a hypothetical era prior to the singularity) – would inevitably be absorbed by the superdense matter near the singularity. This practical horizon is of course somewhat closer to us than the theoretical horizon defined by the emission of a signal exactly at the moment of $\rho = \infty$. Besides, it is different for different kinds of particles – as it must be, for example, for photons and neutrinos since the latter have much higher powers of penetration than the former. But all the differences in the radii of the neutrino, the photon and other horizons turn out to be negligibly small when compared to the horizon radius itself, which amounts to 10–20 billion light years today. The uncertainty in the value is due to the uncertainty in the value of the Hubble constant H.

6   Why no gravitational paradox occurs in relativistic cosmology

The most 'direct and simple' answer to the question posed in the heading would be 'no paradox arises because Einstein's equations, in contrast to Newton's theory which leads to an ambiguity when calculating the force of gravity in the Universe (see § 12 of Chapter 1), yield a unique solution for the uniform matter distribution in the Universe'.

This is certainly true. But the answer in such a form does not clarify the essence of the problem. One would like to elucidate the essential difference between the two theories and their equations that leads to different conclusions.

The author happened to hear various answers to this question, among them even some proposed by physicists, which turned out to be wrong. For example, some people say that the difficulties in Newtonian theory originate from the instantaneous propagation of gravity, which makes any galaxy feel the attraction of all the rest of the matter in the Universe. Einstein's theory, on the contrary, forbids faster-than-light propagation of gravity and, as a result, any galaxy experiences the gravitational action of the matter within its horizon only, which eliminates all the difficulties connected with infinite extensions.

Such an explanation could not be correct. In Einstein's theory only variations of the gravitational field (gravitational ripple) propagate with finite velocity. The quasi-static gravitational field itself, generated by masses (the field which in the limit of Newtonian mechanics obeys the inverse square law), exists in Einstein's theory from the very beginning, is not propagated anywhere and extends to infinity (similar to Newton's). One cannot, for example, at some initial time place a massive star (or a star system) in an empty space with a gravitational field vanishing suddenly at some finite distance from the star and watch how Einstein's theory describes the propagation of this field in space at later times. The same Einstein equations require that, even at the initial moment of time, the gravitational field of an isolated body extends infinitely in space. Mathematically this is reflected in the fact that some of Einstein's field equations were 'conceived' to regulate the conditions on the initial spatial hypersurface. Only changes in the strength of the gravitational field – according to Einstein's theory, and differing in this respect from Newton's theory – propagate with the speed of light.

Thus, to say that in Einstein's theory distant masses exert no gravitational influence and because of that no gravitational paradox arises, would be to give, at least, a misleading answer.

What is the actual reason then?

The true answer lies much deeper, in the very essence of the difference in the approaches to the problem by the two theories.

Using Newtonian theory one first has to calculate the forces of gravity in absolute Newtonian space, and then to determine the motions of galaxies under the action of these forces.

The forces are vector quantities. At any point in space they must be directed somewhere. But in a homogeneous isotropic Universe all directions are equal, while the force must give a preference to some definite direction at every point. Here again one sees the self-contradictory nature of the cosmological problem as formulated in the framework of Newton's theory.

Einstein's theory assumes no absolute space and is not aimed at the calculation of an absolute force of gravity. From Einstein's gravitational equations one finds directly the relative accelerations and relative velocities of galaxies and the geometry of space, i.e. the quantities that in principle can be observed and measured. For this reason alone the conditions of homogeneity and isotropy enable one to solve Einstein's equations without paradoxes. Note that even in Newton's theory, when one seeks a solution to the cosmological problem satisfying the condition that the relative accelerations do not depend on position and direction and are determined exclusively by the separations of galaxies, no paradox arises. One arrives unambiguously at eq. (10) of Chapter 1. Ambiguity appears when one tries to determine the absolute forces of gravity acting in absolute space (and other similar quantities). But absolute forces of gravity are unobservable fictitious quantities. When a body moves freely, along a geodesic, it is weightless and feels no forces at all. Gravitational forces appear when something interferes with its free motion along a geodesic. But these forces depend on a specific obstacle (in mathematical terms, on the reference frame associated with the obstacle). On the Earth, for example, the surface prevents bodies from falling freely downwards and this defines the force. But in a jet-driven rocket the force pressing the pilot down into his seat would be quite different. In general relativity, by the way, the forces of gravity are usually called not merely gravitational forces, but gravitationally inertial forces – their dependence on a particular choice of reference frame being emphasized in this way. Clearly, for an isolated gravitating body there exists a preferred reference frame associated with this body. The forces of gravity calculated in this frame objectively convey the accelerations arising from the gravitational attraction of the body.

In cosmology, however, in a homogeneous Universe, no preferred location exists; the galaxies move freely, not meeting any obstacles (in the same sense as a satellite moves freely around the Earth). Gravitationally inertial forces do not appear in the reference frame associated with the galaxies.[†] In this frame they are objectively absent. The only observable quantities in cosmology are relative velocities and relative accelerations of galaxies. And general relativity theory tells us in a clear and straightforward way how to determine these quantities without any paradoxes.

[†] We are considering now the overall average gravitational field of the entire galactic system of the Universe and neglect possible local enhancements in gravity due to the chance presence of nearby bodies or systems of bodies. Of course, the gravity of these bodies (a nearby galaxy or a cluster of galaxies, for example) shows up in the reference frame being discussed, but it is of no interest when analysing the gravitational paradox and only the total field of all the galaxies in the Universe matters.

3

The hot Universe

1 Physical processes in the expanding Universe

So far we have discussed mostly the mechanical and geometrical properties of the Universe, having said almost nothing as to the physical processes occurring in the expanding Universe. This and the subsequent chapters of the book are devoted to the physics of the expanding Universe.

After World War II and especially over the last two decades astrophysicists have developed a deep interest in the physical processes occurring at different stages of the evolution of the expanding Universe. This was instigated by outstanding achievements in theoretical physics and new methods of astronomical observations – such as radio astronomy, new large telescopes, modern soft-ware, electronics, instruments mounted on rockets, satellites and spacecraft (bringing, in particular, X-ray astronomy into being) – which provided powerful means of exploration of these processes.

As we shall see below, the physical processes at different stages of the expansion of the Universe are quite different in nature and in the consequences they lead to, and it is no wonder, because the conditions that prevailed, say, at the beginning of the expansion and those prevailing today are very different indeed.

In this chapter we consider the processes that occurred at the initial stage of the expansion of the Universe, long before the first galaxies and other individual bodies began to form. The processes that led to the formation of the present structure of the Universe are discussed in the next chapter.

To calculate the physical processes, in the first place one has to know how the expansion of the Universe proceeds. Friedmann's model provides the law of expansion for the homogeneous isotropic Universe. Observations confirm that today our Universe is expanding isotropically and the matter is distributed uniformly on a large scale to a high degree of accuracy. But has it always been like this in the past? Perhaps the earliest stages of expansion proceeded in quite a different way and cannot be described with Friedmann's model, and only later the large-scale inhomogeneities smoothed out and the expansion of the Universe stabilized with a pattern almost exactly conforming to Friedmann's model.

Various theoreticians have succeeded in the construction of a number of different cosmological models in which the Universe expanded highly anisotropically at the very beginning, but approached the Hubble-law expansion afterwards and are indistinguishable from Friedmann's model at the present epoch. But what occurred in the real Universe and what were the physical processes near the singularity? One of the major reasons that this is important is due to the fact that the processes near the singularity had a direct bearing on the subsequent conditions in the evolving Universe under which the present-day picture that we observe – with galaxies, stars, planets and life – has emerged. Observation of just the contemporary expansion pattern cannot disclose the past evolution of the Universe.

Fortunately, observations of the cosmological expansion and of the average matter density in the Universe are not the only means of checking cosmological theories. In different models of the Universe various physical processes proceed in different ways with different consequences. For example, different rates of expansion of the Universe near the singularity affect the physical processes at that time and result in the different chemical compositions of matter from which galaxies and stars (and we ourselves) were later formed, as well as in other different consequences. We shall discuss this in more detail below.

From this we infer that in order to determine how the expansion proceeded near the singularity and what processes occurred then, one has to perform a series of calculations under different assumptions as to the expansion model, and the state and composition of matter in the Universe, and then to compare the results of calculations with the observational data. In this way we are able to elicit the correct assumptions and gradually, step by step, to reconstruct the entire picture of the past evolution of the Universe.

2   Ten billion years BC

As was already mentioned in the previous section, this chapter deals with the physical processes that occurred at the very beginning of the expansion of the Universe, i.e. 10–20 billion years ago. The problem of the reconstruction of the history of the Universe is certainly no less difficult than that confronting archaeologists and historians when they try to reconstruct the history of a long since vanished ancient civilization from scarce fossil debris, though their methods are quite different. According to a scheme outlined at the end of the previous section, we have to try different versions of the expansion law near the singularity on the one hand, and the state of the matter on the other, and then to analyse the ensuing conclusions. Clearly, with every trial we must change one parameter only, to be sure that the ensuing changes in the conclusions are due to this specific parameter and not to any other.

We begin by considering various possible physical conditions at the start of the expansion, having fixed the law of expansion (i.e. the 'mechanics' of the matter). Assume that today – if we consider large enough volumes of space – as well as in the remote past, the expansion of the Universe proceeds and proceeded exactly as Friedmann's theory describes. The other possible versions of the initial stage of the expansion are discussed in §7 of this chapter. Friedmann's solution, when extrapolated to the past, leads formally to the initial state of infinite density. If the matter density in the past was really highly homogeneous and its motion was really highly isotropic, Friedmann's solution can be applied back to the times of huge densities, until very close to the singularity where one encounters a situation that cannot be described within the framework of general relativity theory but requires for its treatment another, more general theory.

Why does the need for such a theory arise? The point is that modern quantum theory predicts quantum gravitational effects in the immediate vicinity of the singularity. Einstein's gravitational theory is a non-quantum theory, and it cannot describe the effects resulting from quantization on the scale of the entire Universe. How can one ascertain the limits beyond which quantum gravitational effects must be accounted for? For this, dimensionality considerations can be of some help. Such considerations make it possible to estimate approximately the range of parameters within which one or another process is important, even when one has no detailed theory of the process at hand and knows only its most general characteristics – such as the universal constants that must enter the detailed theory, for example. In our case the dimensionality considerations can be applied as follows.

Our goal is to establish the radius of curvature of the spacetime at which the following phenomena play crucial roles: gravity, quanta and relativity (high speeds). The effects of gravity are represented by the gravitational constant G, the quantum effects by Planck's constant $\hbar$, and the relativistic effects by the speed of light c. We are interested in the curvature radius, i.e. in a quantity which is measured in units of length and is characterized by the three different types of physical phenomena. In this case the dimensionality considerations require that we take the product of all three constants raised to certain powers to yield a quantity with the units of length. Thus, we arrive at

$$r_{Pl} = (G\hbar/c^3)^{\frac{1}{2}} \approx 10^{-33} \text{ cm.} \tag{1}$$

The above distance r_{Pl} is called the Planck length; it is infinitesimally small; $r_{Pl} \approx 10^{-33}$ cm.

So, the quantum gravitational effects can no longer be ignored once the radius of curvature of the spacetime becomes comparable with r_{Pl}. Under such conditions Einstein's gravitational theory breaks down. Now one can easily calculate at what time after the start of the expansion the radius of curvature was equal to r_{Pl} and what the matter density was then. To estimate the first quantity, divide r_{Pl} by the speed of light c to arrive at $t_{Pl} \approx 10^{-43}$ s, while for the second we obtain the formidable value of

$$\rho_{Pl} \approx 10^{93} \text{ g cm}^{-3}. \tag{2}$$

This density exceeds the present mean matter density in the Universe by more than 120 orders of magnitude, and the density of matter in atomic nuclei by almost 80 orders of magnitude!

Thus, to describe the processes at times $t < 10^{-43}$ s (here and below the time is measured from the singular state), one has to construct a new theory. We return to this problem again in Chapter 4. But is Friedmann's solution legitimate at more moderate densities?

In any case, the general physical laws can be regarded as firmly established up to the nuclear matter density, $\rho_{nucl} \approx 10^{14}$ g cm^{-3}.

Therefore, we assume that Friedmann's solution can be extrapolated back in time at least to the densities of the order of $\rho_{nucl} \approx 10^{14}$ g cm^{-3}. Let us take a closer look at the conclusions ensuing this assumption and compare them with the observations. Clearly, at the epoch of such high density neither stars nor any other celestial bodies existed. (Concerning the formation of individual objects see the next chapter. The so-called primordial black holes are discussed in the last chapter of this book.) The expanding matter was, most probably, almost perfectly homogeneous. What processes occurred in it at a density, say, of the order of ρ_{nucl}? At first glance these processes seem to depend on the conditions that prevailed at still earlier times, of which we known nothing, and nothing of interest therefore can be said about the processes. Fortunately, this is not so. The fact is that at superhigh densities all the processes of interconversions of particles proceed quite rapidly – much more rapidly than the matter density changes in the course of the expansion, i.e. than the conditions under which the reactions occur change. So, all the possible reactions have enough time to settle at the equilibrium level before the physical conditions undergo a significant change. In other words, the matter must be in a state of thermodynamical equilibrium. This means that the composition of matter at any time is completely determined by the conditions at that particular time – by the values of such quantities as matter density, temperature, etc. – and does not depend on its previous history. It is of no surprise, since all reactions are quick enough to bring the matter composition into accord with the existing conditions – irrespective of what had been happening before – in no time at all as compared to the expansion rate. The calculations show that this is exactly the case for Friedmann's model.

The state of equilibrium at a density of 10^{14} g cm^{-3} can be adequately described within the framework of the existing theory. Our complete ignorance as to the properties of matter at a density of, say, 10^{30} g cm^{-3} (at which the behaviour of unknown particles may be governed by unknown laws of nature) cannot prevent us from calculating the equilibrium at $\rho = 10^{14}$ g cm^{-3}. (Recent progress in the theory of elementary particles has led to the conclusion that probably the so-called phase transitions played an important role at the earliest stages of the expansion, and the preceding history is not in fact irrelevant. However, we are not able to dwell on these specific points here – the more so since no complete theory actually exists as yet. See

D.A. Kirzhnitz, *Uspekhi Fiz. Nauk*, **125,** 169 (1978), on this.) Of course, in order to calculate what occurs in this matter, what its composition resulting from fast nuclear reactions will be, it is not sufficient to know just the density of matter – one has to specify also the values of temperature and some other parameters. In the most general case there are several such numerical parameters. From physical considerations we infer that at least two extra numbers must be specified in addition to the density of matter. (Strictly speaking, actually more than two numbers need to be specified, but we are interested here in the two most important of them.) The first number characterizes, roughly speaking, the heat content of matter. This is the famous entropy, of which physicists are so fond. In our case the entropy of matter has a quite clear meaning. The hotter the matter is, the more electromagnetic quanta (photons) there are present in it (and the greater the energy of each quantum). As a result, the number of photons per heavy particle (baryon) adequately represents the heat content of the matter – its entropy. The second number is the so-called lepton charge. For our purpose it is sufficient to say that the lepton charge of the Universe represents the difference between the numbers of neutrinos and anti-neutrinos per baryon plus the difference between the numbers of electrons and positrons per baryon of the Universe. (Strictly speaking, at least three different lepton charges exist – the electronic, muonic and tau-leptonic ones, corresponding to electron, muon and (as yet hypothetical) tau neutrinos. Only the electronic lepton charge and electron neutrinos are discussed here.)

Once specified at some fixed moment of time, both numbers remain almost constant and do not change in the course of the expansion of the Universe. This is their principal advantage over other possible physical parameters (such as the temperature, for example) that could characterize the state of matter equally well, but which strongly vary in the course of the expansion.

So, to calculate the physical processes (nuclear reactions) at the beginning of the expansion within the framework of Friedmann's model, one has to specify the values of two numbers – the entropy S and the lepton charge L. Having calculated the outcome of nuclear reactions, one can predict the chemical composition of the matter from which galaxies, stars and interstellar gas form. Having compared these predictions with the observational data, one can elucidate the true conditions in the past.

3 A hot or cold beginning?

Thus, the problem of the initial conditions reduces to a non-contradictory choice of the two numbers L and S.

The simplest possible choice that immediately comes to mind would be $S=0$ and $L=0$. Such an assumption corresponds to the idea put forward in the 1930s, when no theory of superdense matter existed, that all the matter in the Universe once consisted of cold neutrons only. With no theory of nuclear matter available, the physicists at that time could perform no reliable calculations of nuclear reactions under this assumption.

As we now know, such a model for the initial matter composition contradicts the observations. The fact is that, as the expansion proceeds, every neutron n decays into a proton p, an electron e^- and an antineutrino $\bar{v}$:

$$n \rightarrow p + e^- + \bar{v} \tag{3}$$

A proton created in this way encounters another neutron and combines with it, to form a deuterium nucleus D. The chain of nuclear fusion reactions ends when α-particles – the nuclei of helium – form. In this way all the matter ultimately converts into helium. This conclusion is in strong disagreement with the observations. It is well known that the young stars and the interstellar gas are predominantly composed of hydrogen rather than of helium.

Thus, the observational data compel us to reject the cold neutron hypothesis (with the entropy $S=0$) of the primeval matter.

Another possible set of initial conditions was proposed by the physicist G. Gamov and his coauthors in 1940s and '50s. It was the so-called 'hot' version of the initial stage of the expansion of the Universe. It assumed that the temperature at the initial stage was rather high, i.e. the entropy of the Universe was quite large ($S \gg 1$). When the entropy S is large, the specific value of the lepton charge L becomes insignificant (unless it is extremely high) – the outcome of nuclear reactions is almost independent of L (this statement will be expounded below). Let us put $L=0$ then. The main goal pursued by the authors of the hot Universe hypothesis was, having calculated the nuclear reactions at the initial stage of the cosmological expansion, to explain the observed abundances of various chemical elements and their isotopes.

Why was it assumed at that time that all the chemical elements were created at the very beginning of the expansion of the Universe? The reason for this stemmed from the common belief in the '40s that the time that had elapsed since the start of the expansion was 2–4 billion years (instead of 10–20 billion years according to modern estimates). As we now know, this delusion originated from the underestimated values of galactic distances and, as a consequence, from the overestimated value of the Hubble constant. Comparing this time $(2–4) \times 10^9$ years with the age of the Earth $(4–6) \times 10^9$ years, the authors were forced to assume that even the Earth and other planets (to say nothing of the Sun and the stars) had condensed from the primeval matter, and therefore all chemical elements must have been created at an early stage of the expansion of the Universe since any other alternative would require much longer times.

We now know that the expansion time is $(10–20) \times 10^9$ years. The Earth formed not from the primeval matter, but from the material that had passed through a nucleosynthesis stage in star interiors. The theory of nucleosynthesis in stars was quite successful in explaining the ratios of cosmic element abundances, assuming that the first stars had formed just from pure hydrogen, or from a mixture of hydrogen and helium. So, there is no longer any need for the early stage of the expansion to account for the origin of all the chemical elements – including the heavy ones such as iron, lead, etc.

On the other hand, many researchers argued that the abundance of helium in stars and in the gas within our galaxy by far exceeded the amount that could be accounted for purely by nucleosynthesis in the stars. (For more details on this see §7 of Chapter 3.) Hence, the synthesis of helium must have occurred at the beginning of the expansion of the Universe. But even today the major constitutent of matter in the Universe is still hydrogen. Why then was not all the matter in the hot Universe converted into helium, as was the case in the cold neutron model ($S = 0$, $L = 0$)?

This is entirely due to the fact that the matter was hot. In a hot matter with $S \gg 1$ many energetic photons are present. These photons break down the deuterium nuclei formed by the fusion of protons with neutrons and cut at the very beginning the sequence of nuclear reactions resulting in the synthesis of helium. When the expanding Universe becomes sufficiently cool (when its temperature drops below a billion degrees), a certain amount of deuterium survives and opens up the way to the formation of helium. We discuss this process in greater detail in §6 of Chapter 3.

The theory of the hot Universe predicts a definite value for the abundance of helium in the pre-stellar matter. As we shall demonstrate later, the abundance of helium must be some 30% by mass. Finally, as the Soviet physicist Ya.B. Zel'dovich pointed out in the early 1960s, there is no need in fact to assume the Universe to be hot in order to avoid the ultimate conversion of all the matter into helium. One can remain within the framework of the cold model and demand only that the lepton charge is not zero.

In this last model the matter at the initial stage of the cosmological expansion is supposed to have consisted of protons, electrons and neutrinos in equal numbers with the lepton charge $L = 2$ and the entropy $S = 0$. Equal numbers of electrons and protons are required for the matter to be electrically neutral.

If no neutrinos were present, the protons at high densities would have captured the electrons and would have all been converted into neutrons and neutrinos:

$$p + e^- \rightarrow n + v. \tag{4}$$

As we saw earlier (when discussing the case $S = 0$, $L = 0$), the cold model with all the matter in the form of neutrons is no good since it results in 100% helium, which contradicts the observational data.

The point of the introduction of extra neutrinos into the cold model is that protons cannot convert into neutrons at high densities, in accordance with eq. (4), if neutrinos are already there. These pre-existing neutrinos inhibit the creation of new neutrinos and the process is forbidden.

As the expansion proceeds further, protons do not undergo any conversion. Thus, when neutrinos are present, cold matter transforms into almost pure hydrogen. All the remaining elements, according to this hypothesis, are formed in stars.

Initially, the hypotheses of hot and cold universes were intimately related to the attempts to find a complete explanation of the element abundances in the pre-stellar matter. All the efforts to find decisive evidence in favour of one or another model were initially concentrated on the analysis of observational data concerning the abundances of chemical elements. But the observations involved – and especially their analysis – are very complicated and depend on many additional assumptions. At the same time, the theory of the hot Universe yields a most important observational prediction which is a direct consequence of the high entropy of the matter. It is the prediction of the relict electromagnetic radiation, which must have remained from that distant epoch when the matter was dense and hot, and which must pervade the entire Universe today.

As the cosmological expansion continues, the temperature of the matter drops, and so does the temperature of the radiation, but, nonetheless, some portion of the electromagnetic energy, characterized by the temperature (ranging from a fraction of a degree in some models to 20–30 K in others), must reach today's observers.

Such radiation, that must have remained from the most ancient times in the history of the Universe (if the Universe was indeed hot) has been called the relict radiation. (This name was first proposed by the Soviet astrophysicist I.S. Shklovsky.) Electromagnetic radiation with this low a temperature will appear as the radio waves in the centimetre and millimetre (microwave) wavelength bands. Hence, a decisive experiment enabling one to choose between the hot and cold models of the Universe would be to search for this radiation. If it is present, the Universe was hot, if not, the Universe was cold.

4   The discovery of the relict radiation

The first theoretical estimates of the expected temperature of the relict radiation were given by G. Gamov and R.A. Alpher in their papers published in the 1950s. Their value was some 5 K. But is it possible to detect this radiation against the background of the electromagnetic emission from stars and radio galaxies? In the paper of 1964 by the Soviet astrophysicist A.G. Doroshkevich and the author of this book, the amount by which the intensity of the relict radiation (assuming there is such a radiation, of course) at centimetre wavelengths should exceed the intensity of microwaves emitted by radio galaxies and other sources was first calculated explicitly. In this way the possibility was established to perform a key experiment, providing unambiguous evidence that could decide between the hot and cold Universe models. This theoretical paper, however, went unnoticed by the observers at that time.

The relict radiation was discovered quite accidentally in 1965 by A.A. Penzias and R.W. Wilson of the Bell Laboratories, when calibrating the horn radio antenna devised to track the satellite 'Echo'. They found a weak excessive radio noise of cosmic origin, which was independent of the direction in which the antenna was pointing. R.H. Dicke, P.J.E. Peebles, P.J. Roll and D.T. Wilkinson immediately interpreted the data obtained by Penzias and Wilson as cosmological evidence for the hot model of the Universe. At that time Dicke and his collaborators were building an antenna of their own to look for the radio background that might be created by the relict radiation at a wavelength of 3 cm. The first measurements by Penzias and Wilson were carried out on 7.35 cm radio waves. They demonstrated that the temperature of the radiation discovered was about 3 degrees on the absolute Kelvin scale. Many other independent measurements of the relict radiation at wavelengths from tens of centimetres to fractions of a millimetre have been performed in subsequent years.

These observations clearly demonstrated that the relict radiation has an equilibrium spectrum in complete agreement with the prediction of the hot Universe model. Its spectral shape neatly conforms to the Planck curve for the equilibrium radiation with a temperature of 2.9 K. Fig. 21 shows the total spectrum of the cosmic electromagnetic radiation over a wide range from metre radio waves to ultraviolet wavelengths. (This is, of course, the average radiative spectrum for the entire Universe, which is found far from stars and other individual sources.)

In the metre waveband the most powerful emitters are the radio-galaxies discussed in §3 of Chapter 1. These galaxies are crammed with magnetic fields and energetic electrons. The fast electrons moving through regions of magnetic field emit radio waves. The visible light is emitted predominantly by stars; the most prolific source of infrared radiation seems to be the dust heated by the stellar light. Other sources of infrared emission are also possible. Between these two bands – the radio waves and the visible light (together with the infrared region) – there is a spectral region where the relict radiation dominates.

It is interesting to note that the first time one of the effects of the relict radiation was actually discovered was as early as in 1941. It was then that the astrophysicist A. McKellar noticed that the radicals of cyanogen in the interstellar gas were often observed in rotationally excited states, corresponding to an excitation temperature of ~ 2.3 K. No satisfactory explanation of the excitation mechanism was found at that time. After the relict radiation had been discovered, I.S. Shklovsky, and independently G.B. Field, N.J. Woolf, P. Thaddeus and others quickly recognized it as the exciting agent for cyanogen molecules. From the observations of the molecular lines of CN an independent estimate for the temperature of the relict radiation at the wavelength $\lambda \approx 0.26$ cm has been derived.

When the intensity of the relict radiation at one and the same wavelength was measured in different directions, no differences, within the experimental errors, were found. Experimental errors do not exceed a few tenths of a per cent. This is good evidence that the expansion of the Universe is highly isotropic now and has been so since very remote epochs in the past, when the matter was thousands of times denser than today. Really, the Universe today is practically transparent to the relict radiation which comes to us from enormous

Fig. 21. The spectrum of the electromagnetic radiation in the Universe in the vicinity of the relict radiation wavelengths.

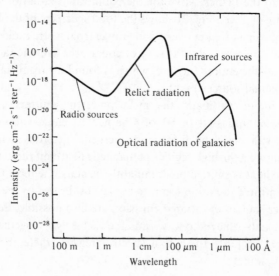

distances. More details on that are given in §8 of Chapter 3. Only for the earliest stages of expansion does there remain the possibility that the Universe did not obey the laws of Friedmann's model!

Unlike star light or radio waves emitted by radio-galaxies, the relict radiation did not originate in some particular kind of source. The relict radiation existed from the very beginning of the expansion of the Universe. It pervaded the whole of the hot matter that expanded from the singular state.

On estimating the total energy density of the relict radiation today, we obtain a value which exceeds by a factor of 30 the energy density of the radiation emitted by stars, radio-galaxies and all other sources added together. One can readily evaluate the number of relict photons per cubic centimetre of the Universe. This number turns out to be

$$N \approx 500 \text{ photons per cm}^3. \tag{5}$$

Remember that the average density of the ordinary matter in the Universe is about 10^{-30} g cm^{-3} (see Chapter 1). This implies that, if all the matter in the Universe were spread evenly throughout space, every cubic metre would contain just one atom (recall that atoms of hydrogen – the most abundant element in the Universe – have a mass of about 10^{-24} g). At the same time, every cubic metre of space contains about a billion relict photons.

Thus, the quanta of electromagnetic waves – particles of a rather peculiar kind – are much more abundant in nature than the particles of ordinary matter. As was discussed in §2 of Chapter 3, the ratio of the number of electromagnetic quanta to the number of heavy particles is a measure of the entropy of the Universe. In our case this ratio is

$$S = 10^9/1 = 10^9. \tag{6}$$

(Note that the number of photons per unit volume is known from observations quite well, while the density of ordinary matter – as we saw in Chapter 1 – has been much more poorly determined. For this reason the ratio given by (6) may change when a more accurate value of the matter density becomes available. Thus, for the critical density ρ_{crit} we have $S = 10^8$.)

So, the Universe has a very high entropy. As we have already mentioned, the ratio given by (6) is practically unchanged in the course of the evolution of the Universe.

Without any doubt, the discovery of the relict radiation was an outstanding achievement in contemporary science. It implies that the Universe at the earliest stages of its expansion was hot. The relict radiation had been predicted in the framework of the theory of the expanding Universe and its discovery demonstrated once more the fruitfulness of the approach to the cosmological problem pointed out by A.A. Friedmann.

5   The very first instants

The first sections of this chapter dealt with the physical conditions at the early stages of the cosmological expansion, and nuclear reactions were mentioned that might occur at densities of about 10^{14} g cm^{-3} and somewhat lower.

When studying the processes near the singular state, theoreticians find it convenient to reckon time from the moment when the matter density was theoretically infinite. This moment is conventionally taken to be zero time. Such a convention does not mean of course that there was 'nothing' before zero time – just as the adoption of the midnight as the zero hour of the day does not mean that there was no yesterday – albeit we do not know yet what occurred in the 'yesterday' of our Universe (we shall return to this issue later). At what time from the start of the expansion did the density drop to a value of the order of 10^{14} g cm^{-3}? An approximate value of this time can be estimated without much difficulty. To do this recall the relation between the critical matter density and the Hubble constant

$$H^2 = 8\pi G \rho_{crit}/3 \tag{7}$$

derived in Chapter 1. We know that the time that has elapsed since the start of the expansion of the Universe is of the same order of magnitude as the inverse of the Hubble constant: $t \approx 1/H$. Equation (7) can be therefore rewritten as

$$t \approx (3/8\pi G \rho_{crit})^{\frac{1}{2}}. \tag{8}$$

Eq. (8) applies certainly not only to the present epoch but to any other (and to those in the past, in particular) as well. As we consider earlier and earlier epochs, the value of the critical density goes up in conjunction with smaller values of the time t that has elapsed since the singularity. If the matter density in the Universe were exactly equal to the critical value, the subscript 'crit' in eq. (8) could be omitted and this formula would relate the time t to the actual matter density in the

Universe. Neglecting numerical coefficients of the order of unity, we can rewrite eq. (8) as

$$t \approx (1/G\rho)^{\frac{1}{2}}. \qquad (9)$$

The actual matter density in today's Universe is not in dramatic disagreement with the critical value ρ_{crit}. Consequently, eq. (9) yields a fair approximation to the time t as expressed through the density ρ. As we pass to earlier times, the estimate given by eq. (9) of t through ρ can be shown to become more and more accurate (we shall not do it here). Near the singularity the effect of pressure on the gravitational field and, consequently, on the law of expansion must also be accounted for. But this does not change the estimate given by eq. (9). Thus, eq. (9) provides an order-of-magnitude relationship between the density ρ and the time t that has elapsed since the onset of the expansion.

Now, with the use of eq. (9), we find that the density $\rho \approx 10^{14} \, \text{g cm}^{-3}$ was reached at $t \approx 10^{-4}$ s.

Thus, we are concerned with the very first instants of the expansion of the Universe. As compared to more than ten billion years that separate us from the singularity, this time interval seems to be infinitesimally small. Nevertheless, the laws of physics and – as we shall see later – the observational data give us the tools to treat so short a time interval with confidence.

Of course, we cannot see directly the processes that occurred at that period of time, but, rather, must judge them by their consequences. There is always the danger of missing some crucial point in our reasoning. For example, we almost always assume that the matter at that early period was homogeneous both in density and composition. But we wish to emphasize once more that all such assumptions are liable to observational tests via their after-effects, and even if our present analysis has (as is possible) some hidden flaws, either the theory or the observations will eventually disclose them.

By the way, the assumption that the matter was homogeneous in those first instants of the Universe can also be checked, and research is presently being done in this direction.

To conclude our digression, one more remark is in order. The fact that the matter in the Universe was superdense and hot in the past can be regarded as firmly established and is not liable to any revision. It may seem that the discussion of the physical processes in the first instants is much more speculative, but, as will become evident later, these processes have such a drastic impact on the present conditions that their consequences render much confidence in our basic conclusions concerning these earliest times.

After all these preliminary remarks, we turn to a more systematic discussion of the role the relict radiation played during the past evolution of the Universe.

Today, every cubic centimetre of space contains some 500 quanta of the relict radiation. The mean energy of each quantum is about 10^{-15} erg. So, the energy density of the relict radiation is

$$\varepsilon_{rel} \approx 500 \text{ cm}^{-3} \times 10^{-15} \text{ erg} = 5 \times 10^{-13} \text{ erg cm}^{-3}. \tag{10}$$

According to Einstein's formula $E = mc^2$ (dividing eq. (10) by the square of the speed of light) this energy density its equivalent to the mass density

$$\rho_{rel} \approx 5 \times 10^{-13}/10^{21} = 5 \times 10^{-34} \text{ g cm}^{-3}. \tag{11}$$

Recall that the mean density of ordinary matter in the Universe is some 10^{-30} g cm^{-3}, i.e. about two thousand times greater.

Thus, at the present epoch the mass of the relict radiation is negligibly small compared to the mass of the ordinary matter.

Let us look now at how these quantities change in the course of the expansion of the Universe.

The density of the ordinary matter varies in inverse proportion to the volume, i.e. in inverse proportion to the cube of the distance. When the separations of galaxies increase tenfold, the matter density drops by the factor of a thousand, etc.

It is not so simple with the mass density of the relict radiation. The number density of photons varies with expansion in exactly the same manner as the number density of ordinary particles does, but in contrast to the latter, the energy of each photon also changes. As the Universe expands, the relict photons experience a redshift, i.e. their wavelength increases. Consequently, the energy of each photon, which is defined by Planck's formula $E = h\nu = hc/\lambda$, decreases in the course of the expansion of the Universe (here h is the Planck's constant, ν is the frequency of a photon, λ is its wavelength, and c is the speed of light). If so, the energy density (and, therefore, the mass density) of the relict radiation drops in the course of the expansion not only due to the decreasing number density of photons, but also due to the diminishing energy of every individual photon. As a result, the mass density of the relict radiation varies as the inverse of the distance in the fourth power. When the distance scale increases by a factor of 10, ρ_{rel} drops by a factor of 10 000.

Quite obviously, the portion of the relict radiation to the total matter density can only be smaller in the future.

Take a closer look at what happened in the past. When the distances between the galaxies were 10 times less than now, ρ_{mat} was 1000 times greater, i.e. $\rho_{mat} = 10^{-30} \times 10^3 = 10^{-27}$ g cm^{-3}; at the same time ρ_{rel} was 10 000 times greater, $\rho_{rel} = 5 \times 10^{-34} \times 10^4 = 5 \times 10^{-30}$ g cm^{-3}. So, the ratio ρ_{mat}/ρ_{rel} was 200 instead of the present 2000. Earlier still this ratio was even smaller.

So, at some moment in the past, when the matter density ρ_{mat} was 10^{-20} g cm^{-3}, the density of the relict radiation ρ_{rel} was the same 10^{-20} g cm^{-3}. (The exact value of this density depends, of course, on today's value of ρ, which is rather poorly known.) And what about earlier still? At earlier epochs the mass of the relict radiation exceeded the mass of the ordinary matter. Note that the photons were much more energetic then: at $\rho = 10^{-20}$ g cm^{-3} they were in the visible part of the spectrum rather than at radio frequencies.

Thus, the dominant contribution to the total density of physical matter in the Universe at early stages of its expansion was due to the light and, when analysing these stages, we can neglect for some time the insignificant admixture of particles of ordinary matter – that matter which plays so crucial a role today, of which the stars, the planets and we ourselves are composed.

Let us continue our excursion into the past, towards the singularity. The radiation had enormous temperatures close to the singularity. It will be rather instructive to derive an expression relating the temperature of matter in the Universe at its earliest stages of expansion to the time t since the singularity. For this, rewrite the energy density of the relict radiation in terms of its temperature with the aid of Boltzmann formula $\varepsilon_{rel} = aT^4$, where $a = 7.6 \times 10^{-15}$ erg cm^{-3} K^{-4}. Then, using Einstein's formula $\rho = \varepsilon/c^2$, express the mass density of the relict radiation via its temperature, $\rho \approx aT^4/c^2$. Recalling now that the relict radiation completely dominated in the mass density at early stages of the expansion, substitute $\rho = aT^4/c^2$ into eq. (9) and finally arrive at

$$T\,(\text{K}) \approx 10^{10}/(t\,(\text{s}))^{\frac{1}{2}}. \tag{12}$$

In this formula the time t is measured in seconds and the temperature T in degrees Kelvin. At one second from the start of the expansion, for example, the temperature was $T \approx 10^{10}$ K. At smaller t the temperature was higher still. Under such high temperatures the violent processes of creation and annihilation of elementary particles occurred. As an example of such processes, first of all the creation of electron–positron pairs by the encounters of energetic γ-quanta, and its inverse – annihilation of e^+, e^- pairs into light quanta, – should be mentioned,

$$\gamma + \gamma \rightleftarrows e^+ + e^-. \tag{13}$$

An electron–positron pair can be created at the cost of an energy no less than the sum of rest masses of both particles multiplied by the square of the speed of light – which amounts to some 1000 keV. Hence, this process comes into play at temperatures in excess of 10 billion degrees, when many light quanta with this high an energy appear. Encounters between electrons and positrons may result in the creation of neutrinos and anti-neutrinos; the inverse is also possible,

$$e^+ + e^- \rightleftarrows \nu + \bar{\nu}. \tag{14}$$

At still higher temperatures the creation of other heavier particles becomes possible.

Fig. 22. The temperature history of the Universe at the earliest stages of its expansion and particle–antiparticle pairs that were abundant at different epochs. On the annihilation of electron–positron pairs their energy transformed into the energy of γ-quanta and T_γ became somewhat higher than T_ν.

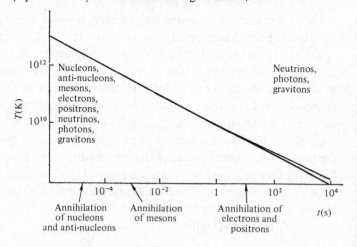

Fig. 22 shows the plot of the temperature variation as calculated from eq. (12). The particles that existed at various epochs of the expansion of the Universe are shown on the figure.

When the temperature was very high, a wide variety of particles (and anti-particles) of different kinds, including heavy particles, existed in roughly equal numbers (see Fig. 22). As the expansion proceeded further, the temperature dropped and the thermal energy of the particles fell below the threshold for the creation of pairs of heavy particles and anti-particles – such as protons and anti-protons, for example. These particles then became extinct.

Later, when the temperature decreased still further, various kinds of mesons became extinct.

A rather important event took place at a time ~ 0.3 s after the start of the expansion. At this era light quanta, electrons, positrons, neutrinos and anti-neutrinos were present.[†]

At high temperatures neutrinos and anti-neutrinos are being continually converted into electron–positron pairs and reformed according to eq. (14).

But neutrinos interact only very weakly with the other forms of matter, and for them even quite dense substances turn out to be transparent. And at a time of 0.3 s the whole of the Universe, including all the electrons and positrons, became transparent to the neutrinos; they ceased to interact with the rest of the matter and decoupled. Subsequently their number has not changed and they have been preserved up to the present time; only the energy of each particle has been decreasing because of the red shift caused by the expansion – in complete analogy with the falling temperature of the electromagnetic radiation.

Thus, beside the relict electromagnetic radiation, relict neutrinos and anti-neutrinos must also be present in today's Universe. If neutrinos have zero rest mass, they, and any other kind of massless relict particles, should possess approximately the same energy today as the relict photons, and their number density should also be about

† So as not to complicate matters, we do not discuss here other sorts of neutrinos – muon neutrinos, recently predicted τ-neutrinos, and other possible types of particles not yet known – all of which would also have been present in the Universe. Nothing has been said of gravitons either. In what follows we discuss briefly the latter and the muonic neutrinos. Remember also that we are not considering so far the insignificant admixture of the ordinary matter.

equal to that of the photons. Note that soon after the decoupling of neutrinos (at a time t of a few seconds) electrons and positrons became extinct. The major portion of their energy was transformed into the energy of photons (see Fig. 22). This is the primary reason that the photon temperature is somewhat higher than that of the neutrinos.

The experimental discovery of the relict neutrinos would be of enormous interest. The Universe has been transparent to them since a fraction of a second from the start of the expansion. If we could detect these elusive particles, we would by able to take a direct glance at that remote era in the past which left its imprint on the spatial and spectral distributions of the relict neutrinos.

Unfortunately, there is no hope of detecting neutrinos of such low energies, as must be typical for the relict particles, with the currently available technique.

It is worth noting in this respect that we are actually witnessing today the creation of neutrino astronomy. We are on the threshold of the discovery of neutrinos born in the nuclear reactions near the centre of the Sun. These neutrinos will enable us to have a direct look into solar interiors because the entire mass of the Sun is absolutely transparent to them. The 'neutrinoscopy' of the Sun may provide us with new valuable information on its internal structure. In a similar fashion, astrophysicists will some day face the problem of the 'neutrinoscopy' of the entire Universe.

As was mentioned in the footnote on page 109, muon neutrinos and anti-neutrinos must also be present in the Universe. The Universe became transparent to these particles at a time $t \approx 0.01$ s, i.e. even earlier than for the electron neutrinos. The discovery of these neutrinos would make it possible to take one step nearer the singularity. But the discovery of cosmological muon neutrinos presents an even more formidable task than the detection of their electron counterparts.

Finally, gravitons may also be present in the Universe. However, the coupling of these (still hypothetical) particles with the matter of the Universe is so weak that they would have ceased to interact with it at a time $t \approx t_{Pl} \approx 10^{-43}$ s, i.e. at a time before which non-quantum cosmology breaks down. It would be very interesting indeed to take so deep a look into the early Universe with the aid of gravitons, but so far this is only a dream.

6 The subsequent five minutes

The first five minutes that followed the first instants of the expansion determined quite a lot of the subsequent history of the Universe. In particular, within these minutes the conditions were stipulated under which the shining stars could appear later – namely, a good store of nuclear fuel had been secured for them. The stars and other heavenly bodies form from what was at that distant epoch an insignificant amount of ordinary matter which we neglected in our discussion of the photons and particle–anti-particle pairs of the previous section.

Let us return now to this minute admixture of ordinary matter 'boiling' during the first fractions of a second in a hot 'soup' of neutrinos and anti-neutrinos, electrons and positrons, and light quanta. The fact is that the processes involving ordinary particles are very sensitive to the conditions that prevailed in the Universe during the first seconds of its expansion. These processes determined the chemical composition of the matter which much later, in an epoch close to modern times, gave rise to galaxies and stars. That is why the chemical composition of stellar material serves as the most sensitive indicator of the physical conditions at the beginning of the cosmo-logical expansion.

Consider the processes involving particles of the ordinary matter. What state was this matter in? First of all, at temperatures in excess of 10 billion degrees no neutral atoms could exist – all the matter was completely ionized and can be referred to as a high temperature plasma. What is more, no composite atomic nuclei could exist at such temperatures. Any composite nucleus would be immediately smashed by the surrounding energetic particles. Hence, the only heavy particles that could survive under those conditions were protons and neutrons. And even they were closely coupled to the 'boiling soup' of energetic electrons, positrons, neutrinos and anti-neutrinos. Inter-actions with these leptons caused the protons and neutrons to undergo continual and rapid conversions into one another,

$$\left. \begin{array}{l} p + e^- \rightleftarrows n + \nu, \\ p + \bar{\nu} \rightleftarrows n + e^+. \end{array} \right\} \tag{15}$$

These reactions maintained a state of equilibrium between the protons and neutrons. When the temperature was high enough,

$T > 10^{11}$ K, equilibrium concentrations of protons and neutrons were about equal.

But as the Universe continued to expand, the relative number of protons increased at the expense of the neutrons, as is shown in Fig. 23. The equality of concentrations was violated because neutrons are somewhat heavier than protons, and the creation of a proton in reactions (15) is energetically more profitable than the converse – hence the greater probability for the formation of a proton rather than a neutron. If reactions (15) continued at the same high rates, in a few tens of seconds almost no neutrons would remain at all. But the rates of these reactions depend strongly on the temperature (they are proportional to the fifth power of T). As the temperature dropped, the reaction rates decreased, and they had practically ceased altogether after the first few seconds of the expansion. The relative abundance of neutrons 'froze' at a value $n_{fr} \approx 0.15$. Later still, when the temperature had fallen to a billion degrees, the formation of the simplest composite nuclei became possible. By this time the thermal energy of photons and other particles was no longer high enough to split the nuclei. All the neutrons available combined with the protons and were converted eventually into helium-4 nuclei:

$$\left.\begin{array}{l} p+n \rightarrow D+\gamma, \\ D+D \begin{array}{l} \nearrow T+p, \\ \searrow He^3+n, \end{array} \\ T+D \rightarrow He^4+n, \\ He^3+n \rightarrow T+p. \end{array}\right\} \tag{16}$$

In addition, very small amounts of helium-3, deuterium and lithium were synthesized. More complex nuclei, in all practicality, did not form under the conditions that prevailed at that epoch. The fact is that heavy elements could be created in appreciable amounts only through binary collisions of the nuclei and particles available in the cosmological plasma. This means that the formation of more complex nuclei could start from a collision of He^4 with a neutron, proton or another nucleus of He^4, but in none of these collisions could a more complex than He^4 nucleus be produced because no stable nuclei with atomic weights 5 and 8 exist!

In addition, the reactions $He^4+He^3 \rightarrow Be^7+\gamma$, or $He^4+T \rightarrow Li^7+\gamma$, were inhibited by the strong electrostatic repulsion of the reacting particles.

For all these reasons, the nucleosynthesis at the early stage of the expansion did not go beyond the lightest elements and stopped completely in about 300 seconds after the singularity, when the temperature fell below a billion degrees and the energy of the particles became too low to initiate nuclear reactions. The elements heavier than helium are being formed in stellar interiors at the present epoch. Stars live sufficiently long, and even not particularly fast reactions have enough time to reach completion. Elements heavier than iron are synthesized in explosive events (in supernovae explosions). The gas that has passed through a nucleosynthesis stage in stars is partially ejected into space either by slow outflows from stellar surfaces or in violent explosions; subsequently it participates in the formation of future generations of stars and other celestial bodies.

Returning to the synthesis of light elements at the beginning of the cosmological expansion, since all neutrons were bound to helium nuclei, one can easily evaluate the total amount of helium created. Every neutron in a helium-4 nucleus is accompanied by exactly one proton; hence, one obtains the fraction, by mass, of helium (it is usually designated by Y) by simply doubling the initial fraction of neutrons,

$$Y \approx 2n_{\mathrm{fr}} \approx 2 \times 0.15 = 0.30. \tag{17}$$

More elaborate calculations give values between 0.25 and 0.30 (depending on the specific value of the mean matter density today).

So, as can be seen from Fig. 23, after about three minutes the

Fig. 23. The variation of proton and neutron abundances with time and the formation of helium. The dashed lines show the variation of the proton and neutron abundances under the assumption that reactions (15) persist as the temperature drops.

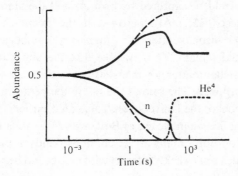

matter consisted of 30% helium nuclei and 70% protons – the nuclei of hydrogen atoms. This chemical composition remained unchanged for the next billion years, until the galaxies and stars began to form and the nucleosynthesis in stellar interiors commenced.

Do the observations confirm the above conclusion about the chemical composition of the pre-stellar matter? We defer the discussion of this issue to § 8 and consider now an important question of how the synthesis of light elements would have proceeded under slightly different conditions at an early stage of the expansion.

7   The synthesis of light elements: a clue to the early Universe

The very fact that the matter from which stars form consists predominantly of hydrogen which was not converted entirely into helium within the first minutes of the Universe (in the latter case the world would appear quite different from what we see now!), is in some sense a happy accident. Only 30% helium formed simply because the reactions given by eq. (15) stopped when the time t was a few seconds, when the fractional abundance of neutrons was 0.15. Had the neutron–proton concentrations frozen not at a time $t \approx 1$ s but at just one tenth of this time, $t \approx 0.1$ s, when the neutron abundance was 0.4, a major fraction (about 80%) of the primordial matter, according to eq. (17), would consist of helium. The time that the freeze occurred is determined by the rates of the reactions of eq. (15). These reactions, in turn, are very sensitive to the rate of the fall in temperature. If the temperature dropped faster than eq. (12) prescribes, interconversion reactions between protons and neutrons would have ceased earlier and more neutrons would have been preserved (see Fig. 24). Later on, all the neutrons would be captured by protons and the amount of helium would exceed 30%. Alternatively, if the freeze occurred later – say, at 100 seconds after the singularity – practically no helium at all would remain. Thus, we see that the abundance of helium in the pre-stellar matter is rather sensitive to the rate at which the temperature dropped. The rate of the fall in temperature in the Friedmannian model of the Universe was calculated earlier – see eq. (12). In calculating this we made use of Boltzmann's formula for the light energy density. But if many as yet unknown kinds of particles exist in nature, which only weakly interact with matter and because of that have not been discovered in laboratories, they also must have been present at the early hot stages of the Universe and their gravity must have affected the rate of the expansion. The result would be a different coefficient in eq. (12). Other processes might also exert their influence on the rate of the temperature change. If, for example, the earliest stages of the expansion departed strongly from Friedmann's model and were highly anisotropic (or matter flowed along some preferred directions), this would also change the rate of the expansion of the Universe. Fig. 25 demonstrates the effect of all these factors on the ultimate abundance of helium in the pre-stellar matter.

One more thing affecting the chemical composition of matter is a possible excess of neutrinos over anti-neutrinos or vice versa – the so-called large lepton charge. Neutrinos and anti-neutrinos interfere with the conversions of protons into neutrons and vice versa. An excess of neutrinos results in a greater abundance of protons, leading to a greater abundance of hydrogen and small abundance of helium in the ultimate chemical composition of the pre-stellar matter. The result of an excessive number of anti-neutrinos would be just the opposite. The plot in Fig. 26 illustrates the effect of a non-zero lepton charge on the final amount of helium.

From Figs. 25 and 26 one sees that there are three most likely values of helium abundance in the pre-stellar matter when arbitrary initial conditions are permitted. These are: (i) practically no helium at all – 0%; (ii) 30% helium, and (iii) 100% helium. Intermediate

Fig. 24. (*a*) Variation of proton and neutron abundances under the assumption that the processes of interconversions between protons and neutrons froze at a time 0.1 s. (*b*) Freezing is assumed to have occurred at a time 100 s.

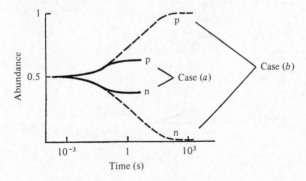

Fig. 25. The abundance of He⁴ in the pre-stellar matter under different conditions at the beginning of the cosmological expansion. The solid line represents the dependence on the number of unknown particles. The abscissa is the ratio of the number of unknown particles to the total number of those already discovered, $\kappa = n_{unknown}/n_{known}$. The dashed line demonstrates the effect of the anisotropic beginning of the expansion of the Universe. The abscissa is the time θ since which the expansion is assumed to have become isotropic. The dotted line is the same as the dashed line but for hypothetical fast motions of matter.

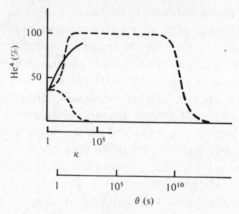

Fig. 26. The abundance of He⁴ in the pre-stellar matter with the excess of either primordial neutrinos or anti-neutrinos. The abscissa is the ratio of the difference between the numbers of neutrinos and anti-neutrinos to the number of heavy particles, $L = (n_\nu - n_{\bar{\nu}})/n_{baryons}$.

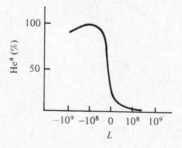

values – though quite possible in principle – require rather special sets of initial conditions out of all the possible ones. The value of 30% He^4 can be obtained in a 'stable' manner only in the framework of Friedmann's model for these early stages of the expansion. In addition, the number of unknown particles and the lepton charge must not be too large.

To conclude this section, we comment on the abundance of deuterium that must have been synthesized at the beginning of the expansion in the 'standard' Friedmannian model of the hot Universe. Its abundance is extremely sensitive to the parameter characterizing the 'hotness' of the Universe – to the value of the entropy S. As we mentioned in §5 of Chapter 3, the uncertainty in the value of S is entirely due to the uncertainty in the mean matter density of the contemporary Universe. Below, we give the values of deuterium abundance (by mass) Z_D as predicted by the theory for two different density values:

$$\left.\begin{array}{ll} \rho_1 = 3 \times 10^{-31} \text{ g cm}^{-3}, & Z_D \approx 10^{-4}, \\ \rho_2 = 10^{-29} \text{ g cm}^{-3}, & Z_D \approx 10^{-12}. \end{array}\right\} \qquad (17)$$

8   The observed light element abundances

What information as to the early evolution of the Universe can be obtained from the chemical composition of the cosmic matter observed today?

First of all, it is quite clear that even if one knows the chemical composition of a cosmic object today, one cannot compare it directly with the predictions of cosmological theory because chemical elements could be created (or destroyed) in the course of the evolution of heavenly bodies – in stars, for example, or by the interaction of cosmic rays with interstellar matter. Hence, one first has to analyse the evolution of chemical element abundances, and only after that compare the results of theory with the observational data.

We begin with the observational data. It is clear from the previous section that, for our purposes, the most interesting element is helium He^4. It is interesting to note that the helium, being so extensively discussed, was first discovered on the Sun (and named after it) in 1868 by the French astronomer P.J. Janssen and the Englishman J.N.

Lockyer. Only afterwards was helium found on the Earth and in the spectra of many other hot cosmic bodies.

Helium has properties that make it very difficult to observe and analyse using normal spectral methods. It cannot be observed in comparatively cool stars. It reveals itself only in hot stars, in huge clouds of gas heated up by ultraviolet star emission, in hot regions of the upper solar atmosphere and in the streams of particles of the solar wind. An indirect estimate of the helium abundance in stars can be obtained by comparing the conclusions of the theory of stellar structure with the observed values of their temperatures and luminosities. In almost all cases the abundance of helium falls within the range $0.26 < Y < 0.32$. The major fraction of all the remaining matter consists of hydrogen. One should not forget that helium is synthesized in ordinary stars and then ejected into the interstellar medium by various mechanisms – from the slow outflow of matter, similar to that of the solar wind, to the violent explosions of supernovae stars.

Could it be that all the observed He^4 was synthesized in the course of the evolution of the first generation of stars and subsequently ejected into space, in this way enriching the interstellar gas from which contemporary objects formed later? A thorough analysis of the problem shows that this is most unlikely. The most convincing evidence against such a possibility was provided by the analysis of the luminosity–temperature diagrams for old star clusters. The appearance of these diagrams depends on the initial abundance of helium in stars and their age. A thorough analysis showed that the initial abundance of He^4 in the oldest stars (and hence the most deficient in metals) of our galaxy was about $Y \approx 0.3$.

The abundance of He^4 in the pre-stellar matter must have been about the same as today, i.e. about $Y \approx 0.3$.

Much scantier information is available on the possible abundance of deuterium in the pre-stellar matter. The difficulty is mainly due to the fact that the primordial deuterium is easily destroyed. At the same time, deuterium is an intermediate product of nuclear processes in stars. A certain amount of deuterium could have been created in star explosions.

Among other methods, use is made of the interstellar molecules HCN and DCN, for example, to estimate the abundance of deuterium in the interstellar gas. It turned out, however, that the relative abundances of these molecules do not reflect the amount of deuterium relative to hydrogen, because the above mentioned molecules have different probabilities of forming due to their different properties. Omitting the details, the observed abundance of deuterium can be estimated as $Z_D \approx 5 \times 10^{-5}$ with a large uncertainty – the true value may be, say, a factor of 10 more. Thus,

$$\left. \begin{array}{l} Y \approx 0.31 \\ Z_D \approx 5 \times 10^{-5} \end{array} \right\} \tag{18}$$

What conclusions can be drawn from these numbers?

First of all, the magnitude of Y is very close to the theoretical value obtained for the simplest and most natural case without many kinds of unknown particles, without an appreciable lepton charge and without significant deviations from Friedmann's law of expansion of the Universe (see the previous section). It must be emphasized that the value of Y is rather insensitive to variations in the contemporary mean matter density in the Universe. Such good agreement between the theory and the observations is strong evidence that the right theoretical model has been adopted.

The amount of deuterium $Z_D \approx 5 \times 10^{-5}$ could have been synthesized at the beginning of the cosmological expansion only if the contemporary matter density ρ has the lowest value possible, $\rho \approx 3 \times 10^{-31}$ g cm^{-3}, which is just the average density of matter making up the galaxies. If this is true, intergalactic gas with an appreciable density (i.e. exceeding considerably the density of the galactic matter and sufficient to make the total matter density ρ in the Universe be equal to, or greater than ρ_{crit}) cannot exist. One should not forget, however, that the true value of Z_D for the pre-stellar matter is still very uncertain.

The above discussion can be summed up as follows. The observations definitely reject all the theoretical models that result in almost 100% He^4 abundance in the pre-stellar matter (see the previous section). At the same time, one can quite confidently conclude that the 'standard' theory described in §6, which yields 30% He^4, has been confirmed by the observations. Hence, we have quite a sound knowledge of what was happening during the first seconds of the expansion of the Universe.

9 After the first million years

In the period 10 s $< t <$ 100 s, annihilation of electron–positron pairs occurred. At earlier epochs when temperatures exceeded 5×10^9 K these pairs filled the Universe in an approximately equal number to the photons (see §5 of Chapter 3). After the temperature dropped below 5×10^9 K, the annihilation of e^+, e^- pairs could no longer be compensated for by the reverse creation processes because the thermal energy of the particles was no longer sufficient to create an e^+, e^- pair. All the energy shared by the electron–positron gas was transformed into the energy of photons. The reader probably remembers (see §5 of Chapter 3) that the matter in the Universe at that time had been long since transparent to neutrinos. Like photons, neutrinos participate in the overall expansion of the Universe and cool down, but, in contrast to the former, they do not interact either with $e^\pm$ pairs or with photons. Hence, they were in no way affected by the energy transfer from $e^\pm$ pairs to photons. As a result of such an energy transfer, the temperature of the photons from that moment became some 40% higher than that of the neutrinos. Subsequent cooling of the photons and neutrinos by the expansion of the Universe proceeded at one and the same pace (at least for quite a long period) and the temperature difference remained constant (see Fig. 22). The annihilation of $e^\pm$ pairs and the synthesis of helium were the last violent events in the history of the Universe, followed by a long quiet period of about a million years during which nothing spectacular has happened.

As the Universe continued to expand, the temperature decreased and the nuclear reactions quickly stopped. The main contributors to the mass in the Universe were the photons, the neutrinos (and probably the hypothetical gravitons), and an admixture of the ionized 'ordinary' matter consisting of 70% protons, and 30% helium, and of electrons neutralizing this high temperature plasma. The neutrinos (and the gravitons as well) can be 'forgotten' for some time since they did not interact with the 'ordinary' matter. The photons, on the contrary, interacted quite effectively with electrons and the matter was opaque to them. This interaction kept the temperatures of the matter and the radiation equal. The epoch under discussion is called the era of photon plasma.

This phase lasted until the temperature fell below 4000 K at $t \approx 10^6$ years.

By that time the protons began to capture electrons very efficiently, and were thus converted into neutral hydrogen atoms. The reverse process, that of ionization, was quickly slowing down at this epoch because the number of photons with energies high enough to ionize hydrogen atoms rapidly decreased, while atomic collisions resulting in ionizations were extremely rare. As a result, the dominant chemical element in the Universe – hydrogen – became neutral. Somewhat earlier helium had been neutralized too.

The epoch at which the cosmic plasma recombined and became neutral played the most important role in the evolution of the Universe. The fact is that the ionized gas was opaque to the relict radiation. After the recombination it became transparent. The relict radiation barely interacts with the neutral gas – it has become decoupled from the matter.

The first important consequence of the decoupling is that the formation of individual celestial bodies becomes possible. This issue is discussed in the next chapter.

The second consequence is closely related to the possibilities of the exploration of the Universe. The moment in time at which the Universe became transparent to the radiation marks the 'practical' observable horizon discussed in § 5 of Chapter 2. When observing the relict radiation today, one virtually 'sees' a sphere bounded by the decoupling era, beyond which the matter was opaque. This sphere lies at a distance corresponding to the redshift $z \approx 1000$.

Following the recombination, matter became 'clear' at the decoupling era quite rapidly, taking about 0.1 of the time interval t_{rec} that elapsed from the singularity to the decoupling itself.

As a concluding remark to this not very long section on the era of photon plasma, we note that this era was not utterly dull: important – though not particularly violent – slow processes, fraught with dramatic consequences for the future evolution of the Universe, occurred at this stage. These were the evolution of small deviations from the homogeneity and isotropy of matter distribution and its motion, and the evolution of the fluctuations of the gravitational field. But this will be the topic of the next chapter.

4

The formation of the structure of the Universe

1   Gravitational instability

One of the most important problems in cosmology is the problem of the stability of the expansion of the homogeneous medium. Even Newton argued that homogeneous matter must eventually gather into one big clump or into a number of individual clumps under the mutual gravitational attraction of particles. In the year of 1692 Newton wrote (in a letter of 10 December to Richard Bentley, quoted in Munitz, 1957):

If the matter of our sun and planets and all the matter of the universe were evenly scattered throughout all the heavens, and every particle had an innate gravity toward all the rest, and the whole space throughout which this matter was scattered, was but finite, the matter on the outside of this space would, by its gravity, tend toward all the matter on the inside and, by consequence, fall down into the middle of the whole space and there compose one great spherical mass. But if the matter was evenly disposed throughout an infinite space, it could never convene into one mass; but some of it would convene into one mass and some into another, so as to make an infinite number of great masses scattered at great distances from one to another throughout all that infinite space. And thus might the sun and fixed stars be formed.

an extraordinarily clear description of the basic idea of gravitational instability. But if all the expanding matter broke into separate clumps at the beginning of the expansion, these objects would never have been able to evolve into galaxies and stars. The fact is that the average matter density in galaxies is some 10^{-24} g cm^{-3}, which, though considerably greater than the mean matter density in the whole of the Universe today, is much less than the matter density at the beginning of the cosmological expansion. If individual objects had formed at some distant epoch close to the singularity, they would necessarily have had a density no less than the mean matter density in the Universe at that epoch, because in the process of their formation they would have had to pass through a stage of contraction. There is also

122

direct observational evidence confirming that matter was quite uniformly distributed in space (at least on scales corresponding to contemporary galaxies, clusters of galaxies and on larger scales) at the era of decoupling. This evidence is the high degree of uniformity in the distribution of the relict radiation intensity over the sky sphere. We have already mentioned in the previous section that, when observing the relict radiation, we actually see the decoupling era. Had the matter distribution at that epoch been inhomogeneous, we would see different intensities of the relict radiation from different directions in the sky (see §4 of Chapter 3 on this). So, no big condensations could have existed at that time. Why then had the matter not formed into separate objects at the very beginning of the expansion, but did so only much later, at an epoch quite close to modern times?

Before concentrating on these evolutionary problems of the Universe, consider in some detail the theory that describes the clumping of homogeneous matter under the action of gravitational forces – the theory of gravitational instability.

What forces resist the action of gravity which tends, according to Newton, to break matter into separate clumps due to the mutual attraction of particles? The elasticity of gas, revealing itself in the increase in pressure accompanying the increase in density, is such a counter-force. Consider what happens when a density enhancement has been created in the gas. Two forces oppose one another: the force of pressure and the force of gravity. The pressure tends to push the gas out of the enhancement region, gravity tends to force it in. Which force wins depends on the size of the density enhancement. For a small enhancement, pressure will dominate, for a large one, gravity will. Let us make some analytical estimates.

Suppose that inside some sphere of radius r the matter density increases by $\Delta\rho$, which results in the increase of pressure by Δp. The excess pressure tends to expand the gas. This force of expansion is directly proportional to the pressure difference Δp and inversely proportional to the size r of the density enhancement – the distance over which the pressure difference Δp develops. The value of the force of expansion acting on every cubic centimetre of gas near the edge of the enhancement is given approximately by

$$F_p \approx \Delta p/r. \tag{1}$$

Clearly, the smaller r is, the more significant the force due to the pressure. Gravity tends to compress the gas. This compressing force can be evaluated from Newton's law of gravity. The force that gravity exerts on every cubic centimetre of gas at the edge of the spherical density enhancement is given by

$$F_g = -GM(\rho + \Delta\rho)/r^2, \tag{2}$$

where M is the mass of the gaseous sphere and ρ is the density of the gas outside the condensation. Substituting the expression $M = \frac{4}{3}\pi r^3(\rho + \Delta\rho)$ for M in eq. (2), we get

$$F_g = -\frac{4}{3}\pi G\rho^2 r - \frac{8}{3}\pi G\rho\Delta\rho r - \frac{4}{3}\pi G(\Delta\rho)^2 r. \tag{3}$$

The largest term in this expression is the first. It does not depend on the amplitude of the perturbation and is none other than the force discussed in Chapter 1 (see §2). It decelerates the overall cosmological expansion of all the matter – inside the perturbation volume as well as outside it – and, since it does not violate the uniform matter distribution, is of no significance to the analysis of the perturbation growth. The evolution of the density enhancement is determined by that part of the gravitational force that depends on the value of the excess density $\Delta\rho$. In other words, we are not interested in the whole of the gravitational force given by (3) but in its excess $\tilde{F}_g$ over the value that one would obtain for the unperturbed density, when $\Delta\rho = 0$. So we are interested in the second and third terms on the right-hand side of eq. (3). The third term is proportional to the square of the density perturbation $(\Delta\rho)^2$ and is much smaller in magnitude than the second term. Neglecting the third term, we arrive at the following expression for that part of the gravitational force that is relevant here:

$$\tilde{F}_g = -\frac{8}{3}\pi G\rho\Delta\rho r. \tag{4}$$

From this formula we see that the gravitational force $\tilde{F}_g$ amplifying a local inhomogeneity in the homogeneous medium grows with the size of the perturbation r. For small dimensions it is insignificant and the pressure force prevails. The critical size of the density enhancement for which both forces cancel each other out is called the Jeans length. J. Jeans was the first to derive this quantity at the beginning of this century. Equating eqs. (1) and (4), we obtain the following formula for this critical length:

$$\Delta p/r = \frac{8}{3}\pi G\rho\Delta\rho r, \tag{5}$$

$$r_{Jeans} = (\Delta p/\Delta\rho)^{\frac{1}{2}}(1/\frac{8}{3}\pi G\rho)^{\frac{1}{2}} \tag{6}$$

Recall that according to hydrodynamics the quantity $(\Delta p/\Delta\rho)^{\frac{1}{2}}$

is the adiabatic velocity of sound v_s in a gas. Thus, the critical length is

$$r_{\text{Jeans}} \approx v_s (1/\tfrac{8}{3}\pi G\rho)^{\frac{1}{2}}. \tag{7}$$

Let the size of a density enhancement be less than the critical Jeans length, $r < r_{\text{Jeans}}$. In this case the dominant force will be the force of pressure. The gravity can be neglected. The pressure force, pushing the gas out of the enhancement region will inhibit its growth and smooth it out – an isolated body cannot form. The pressure will cause the initial condensation to expand and, under inertia, pass the equilibrium state, when the gas density within the condensation becomes equal to the surrounding matter density, thus turning a local condensation into a local rarefaction. Later, the pressure of the surrounding gas will force this region into a recontraction. In this way the gas will swing back and forth in acoustic oscillations which are gradually damped out due to the viscosity.

If the size of a condensation exceeds the critical length, $r > r_{\text{Jeans}}$, then the gravitational force dominates and the density contrast grows. In conclusion: homogeneous matter can separate into clumps only with sizes in excess of the critical Jeans length. Particular sizes and, consequently, the masses of the clumps into which the homogeneous gas fragments depend on the sizes of the initial small density enhancements which were present in the gas. The word 'small' refers here not to the sizes of the condensations but to their excess density over the mean density of the surrounding gas. The sizes and the magnitudes of small density enhancements depend on the preceding history of the medium. But why, nonetheless, did all the matter not fragment into separate clumps with sizes equal to, or exceeding, the Jeans length at the very beginning of the expansion? The point is that the growth of condensations (usually referred to as perturbations) in the Universe must be considered against the background of the expanding matter. The expansion alters the conditons of perturbation growth – as a result of the change in the value of r_{Jeans} – and determines the rate of growth. In the next section we shall discuss various types of perturbations, and in §3 and subsequent sections, the conditions of perturbation growth. At that point the answer to the above question will become evident.

2   The types of perturbations in homogeneous matter

The departures from the homogeneity of the type discussed in the previous section are called adiabatic, or acoustic perturbations. The term 'adiabatic' arises from the fact that the matter density is perturbed without energy redistribution between the macroscopic volumes of matter, i.e. adiabatically.

Other types of perturbations also possible in the homogeneous matter.

For example, vortex motions of matter can occur, when the fluid velocity is different at different points but varies in such a manner that the matter density remains unperturbed. Examples of such motions are shown in Figs. 27c and 27d. Figs. 27a and 27b give the examples of motions of matter that result in density fluctuations. These latter types of motions occur as the result of adiabatic perturbations discussed above.

Returning to vortex perturbations, we note that since they cause no density variations they do not generate perturbing gravitational

Fig. 27. Velocity fields in various types of perturbations: (a) and (b) are examples of the patterns of motion resulting in the perturbations of matter density; (c) and (d) are vortex motions which do not perturb the density. In all cases only the perturbed velocity component is shown, with the velocity field of the overall expansion of the Universe subtracted.

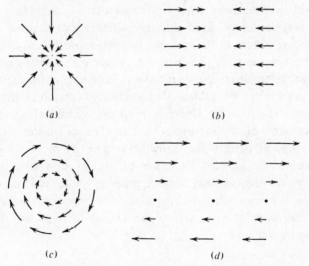

forces resulting in gravitational instability. This is true, however, only for the motions which can be described in the framework of Newton's theory. If the vortexes involve such big masses that Einstein's theory must be used, then non-vanishing perturbing gravitational forces appear. Where do they come from? Recall (see § 1 of Chapter 2) that gravity in Einstein's theory depends not only on the mass distribution, but on the kinetic energy as well (and on the pressure, and magnetic and other fields, etc.). The vortex motions – although not associated with density perturbations – give rise to vortex velocities. And it is the latter that generate relativistic gravitational effects. Of course, when the masses drawn into motion are not large (i.e. the size of the perturbed mass is much less than its gravitational radius) and the velocities are much less than the speed of light, all these relativistic effects are negligibly small.

Finally, there is one more type of perturbation which is very important for the model of the hot Universe.

Consider an early stage of the expansion when the mass density of the relict radiation greatly exceeded that of the ordinary matter. Let the relict radiation be uniformly distributed in space, while condensations and rarefactions are present in the ordinary matter (see Fig. 28). Since only a small amount of ordinary matter is present, the mass, the gravity and the pressure of its particles can be ignored. All the mass and all the pressure is that of the radiation. We have already mentioned that the ionized matter, being opaque to the radiation, is

Fig. 28. Entropy perturbations. The photons (represented by the smaller dots) are distributed uniformly. The ordinary matter (represented by larger dots) is concentrated in randomly spaced clumps.

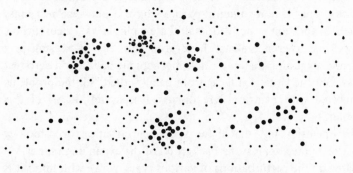

strongly coupled to it. As a result, the initial inhomogeneities neither grow nor dissipate during the whole period preceding the moment of recombination and decoupling – the uniform radiation field remains just 'sprinkled' with some quantity of matter.

Such perturbations have been called the entropy perturbations. This term can be justified as follows. Recall that the entropy (or, rather, the specific entropy) is the ratio of the number density of photons to the number density of protons. The number density of photons in the perturbations being discussed is the same everywhere (the relict radiation is homogeneous), while the matter density varies from place to place – hence the specific entropy also varies from place to place and we do indeed have entropy perturbations.

Up to now we discussed the perturbations of the medium characterized by such quantities as pressure, temperature, etc. A medium that can be described by these quantities is called a fluid. The approximation of a fluid is justified so long as the typical time interval between successive collisions of the particles in the medium is much less than the time scale on which its macroscopic parameters – the density, the mean thermal energy of particles, etc. – change. But there certainly are particles in the Universe for which this condition is violated, which travel in the Universe practically without collisions from the very first instants of the expansion. These are the neutrinos, the hypothetical (not yet discovered) gravitons and perhaps some other as yet unknown particles. We shall see below that the analysis of the perturbation behaviour in such a medium of non-interacting (collisionless) particles may be of much interest because these perturbations might have played a very important role in the formation of the present structure of the Universe. But this could be the case only if non-colliding particles have a non-zero rest mass which, in addition, is not too small. Lately physicists – theoreticians as well as experimentalists – have been with ever increasing persistence suggesting that neutrinos may actually have a non-zero rest mass. But the issue is still far from being settled. All the problems connected with neutrinos are discussed separately in §5 of this chapter.

In conclusion, note that all the types of perturbations in cosmology must be treated against the backgroud of the expanding matter.

How do the perturbations of various types evolve with time? It is quite clear that the answer to this question is intimately linked with the solution to the problem of the origin of the present structure of the Universe – the development of small perturbations into the various types of heavenly bodies and their systems.

If we know how the perturbations evolve with time and what they were like at the start of the expansion of the Universe, we can, using the laws of physics, calculate the whole evolution of the perturbations and can (at least in principle) calculate the entire process of the formation of heavenly bodies. The evolution of all types of small perturbations readily lends itself to a rigorous mathematical treatment. This part of the problem may be considered as solved. Much more difficult is the calculation of the evolution of large inhomogeneities, when the latter fragment into isolated condensations where complicated internal processes commence and the condensations begin to interact with one another. At present, theoreticians are exerting strenuous efforts to achieve some progress in this direction. But nobody knows what the perturbations were at the very beginning of the cosmological expansion, i.e. the initial conditions for the problem under discussion are completely unknown.

Some small departures from the homogeneity ought to have existed, otherwise the matter of the Universe would not have separated into fragments but, rather, would have evolved in the process of the expansion into a cold homogeneous gas, uniformly filling the whole of space. This would be quite unlike what we see at present – clusters of galaxies, galaxies, stars.

The principal obstacle impeding the solution of the problem of the formation of the large-scale structures in the Universe is our ignorance as to the initial conditions.

There seems to be only one possible way to unravel the mystery here, analogous to that used when analysing the physical processes in the first few seconds of the expansion. One has to try different assumptions as to the parameters of the initial perturbations, to calculate their effect and to compare the result with the observational data.

And this is how the theoreticians actually try to cope with the situation. Unfortunately, this new task is much more complicated than the problem of the first few seconds, or even of the first million years of the expansion. There, as the reader will readily recall, one had direct and clear consequences from the processes concerned. These were the helium abundance for the first seconds, and the relict radiation fluctuations for the first millennium. Whenever one chose somewhat different initial conditions, the results changed dramatically.

Unfortunately, this is far from being the case in the problem of galaxy formation. Various types of perturbations could presumably result in more or less the same eventual picture as is observed today. Whether this is true or not, has not yet been firmly established – so long and intricate is the path of evolution from the initial weak inhomogeneity to a galaxy with hundreds of billions of stars, with complex motions inside it, and a complex evolution of the galaxy as a whole.

That is why hot disputes still take place; why no generally accepted theory of galaxy formation yet exists. Impressive work has been done over the last decade by both the theoreticians and the observers. Many complicated processes have been investigated but no consensus has been reached.

Isn't it amazing that we are quite certain about the processes that occurred during the first second of the expansion, and are rather ignorant of the events that have taken place much nearer to our epoch? No, it is not. This is not the first time that science has encountered such a situation. For example, we are quite certain of our knowledge of the internal structure of distant stars and of the nuclear processes that take place inside them, but we are rather uncertain about the interior of our own planet, the Earth, though we live on it! The fact is that the stars consist of a hot gas, and it is much easier to calculate the structure of a gaseous star than the structure of the solid crust and of the semi-melted core of our Earth. The processes inside the Earth are much more diverse than those inside the stars. Analogously, it is comparatively easy to calculate the nuclear reactions in the superdense and superhot matter of the earliest stages of the expansion, and much more difficult to calculate the formation of the heavenly bodies.

The problem of galaxy formation is one of the most urgent problems in modern cosmology.

There are several variations of the theory of galaxy formation, or, one should say, several different hypotheses. Here, we consider only the hypotheses in which the galaxies and stars have originated from condensations in an initially almost homogeneous tenuous expanding medium. Sometimes, different, unorthodox hypotheses are discussed. For one important alternative line of reasoning see 'Works of Academician V.A. Ambarzumyan' (*Collection of Works,* Erevan, 1960).

The hypotheses are classified according to the types of small perturbations with which the formation of galaxies is mainly associated. There are adiabatic, vortex and entropy hypotheses. Finally, the possible importance of the role of non-colliding particles – neutrinos – has been realized recently (see § 5 of this chapter), and the hypotheses of this class may be called the neutrino hypotheses.

In view of the variety of hypotheses we adopt the following strategy. In the next section we discuss the evolution of all the above mentioned types of small perturbations in the gas of colliding particles against the background of the expanding Universe. This problem, as has been already mentioned, can be solved rigorously and unambiguously because the perturbations are small.

The next but one section deals with the more involved question of how the galaxies could form from various types of perturbations after the latter became large.

And finally, in the last section of this chapter, § 5, we speculate on how our ideas concerning galaxy formation may change if the latest claims by some physicists of a non-zero neutrino mass prove to be true. Until the last section we assume that neutrinos have either zero or an absolutely negligible rest mass, and that there are no other massive non-interacting particles.

The reader interested in the problem of the origin of galaxies is referred to the book 'The Origin and Evolution of Galaxies and Stars' edited by S.B. Pikelner, Nauka, Moscow, 1976, and to the review of latest achievements by G.S. Bisnovatyi-Kogan, V.N. Lukash, and I.D. Novikov in the Proceedings of the Fifth European Regional Meeting in Astronomy (Institut d'Astrophysique, Liège, Belgium, 28 July to 1 August, 1980).

3   The evolution of small perturbations in the hot Universe

We begin with a discussion of the evolution of small perturbations appearing as slight condensations and rarefactions of matter, i.e. with adiabatic perturbations.

Intuitively, these seem to be the most likely candidates for the progenitors of galaxies. Historically, as we saw earlier, this idea goes back to Newton.

So we have to apply the idea of the gravitational instability to the hot expanding Universe. Jeans' theory of gravitational instability was first extended to Friedmann's expanding Universe by the Soviet physicist E.M. Lifshitz in the 1940s. In his work the basic principles were laid down for all subsequent studies of the origin of the structure of the Universe.

The theory of galaxy formation in the process of adiabatic perturbation evolution in the hot Universe was most extensively developed in the USSR by Ya.B. Zeldovich's group. Consider how a hot expanding homogeneous medium can divide into separate fragments. The critical Jeans length is determined by the density of the medium and by the speed of sound in it (see eq. (7)). One can calculate its value for any time in the expansion of the Universe. We have already seen that at the earliest stages of the expansion the major contribution to the mass of the expanding medium was due to light. The pressure of the medium was also almost entirely that of light. The light quanta travel with the fastest speed possible in nature. For this reason, the pressure – determined by this speed – is very high. Thus, the velocity of sound is high, amounting to more than half the speed of light. As a consequence, the Jeans length – which we are interested in – is quite large. It turns out to be about equal to the observable horizon (see §5 of Chapter 2), i.e. to the distance covered by light since the beginning of the expansion up to the time under consideration.

One can easily verify the above conclusion having noticed that the velocity of sound in eq. (7) at the early stages of the expansion is approximately that of light, $v_s \approx c$, while the quantity $[3/(8\pi G\rho)]^{\frac{1}{2}}$ represents approximately, according to Friedmann's theory, the expansion time that has elapsed since the singularity up to the moment when the matter density became equal to ρ (see eqs. (8) and (9) in §5 of Chapter 3). As a result, we have

$$r_{\text{Jeans}} \approx ct \approx r_{\text{hor}}. \tag{8}$$

Thus, the Jeans length grows with the expansion of the Universe in step with the horizon and encompasses an ever increasing mass.

The above conclusion about the approximate coincidence between the two scales r_{Jeans} and r_{hor} during the entire radiation dominated era is very important. As we shall see in a moment, it means that under such conditions the growth of small primordial perturbations cannot culminate in the fragmentation of the medium. We shall be discussing volumes whose sizes exceed the critical radius (see § 12 of Chapter 1) beyond which relativistic effects must be taken into account. In our explanations we are forced to use somewhat vague concepts and approximate analogies because the level of this book does not permit us to employ the rigorous, rather complicated machinery of relativistic theory.

Let the homogeneous expanding medium of the Universe be rippled with small random density enhancements. Only condensations with sizes in excess of the Jeans length – i.e. greater than the observable horizon in this case – are able to grow. The gravitational forces tend to increase the density inside these condensations. Consider one such condensation. Since its size exceeds r_{Jeans}, the pressure gradient, as compared to the gravitational forces, is negligible. Under the action of the excess gravitational pull due to the density enhancement, the edge of the condensation† moves toward its centre relative to the unperturbed matter (which continues to expand on its original course). Clearly, the edge cannot move with respect to the unperturbed matter faster than light – in fact its relative velocity will be much less than the speed of light. This means that, even if it began at the start of the expansion of the Universe, the edge of the condensation would cover a distance that was only a small fraction of the size of the condensation – just because the latter is assumed to be larger than the observable horizon, and only light would be able to cover that great a distance. So, since the decrease Δr in the size of the condensation can only be small compared with the size r itself, the relative increase in the density $\Delta\rho/\rho$ (which must be of the same order of magnitude as $\Delta r/r$) can never become comparable to 1, i.e. can never become so large as to stipulate the isolation of our condensation into an individual object.

In the most extreme case, when the gravitational field of a perturbation is very strong (more specifically, in terms of relativistic

† The term 'edge of the condensation' is an idealization enabling us to simplify the argument. Of course, in reality no condensation would have sharp edges: the density would smoothly drop from its central regions to the outskirts and merge with the surrounding medium.

theory, when space is substantially curved on a scale of the size of the perturbation), there is just enough time for a condensation to grow and become large (i.e. for the matter density in the condensation to become large compared to the average density of the surrounding medium) and begin to contract at a time when its size becomes comparable to the horizon (but not earlier!). But in this case the size of the condensation is of the order of its gravitational radius (see § 12 of Chapter 1) and a so-called primordial black hole forms. We discuss primordial black holes in greater detail in § 7 of Chapter 5.

Here we just emphasize once more that this is possible only when the curvature of space (i.e. the gravitational field) in the region of the perturbation is substantially greater than that of the surrounding medium.

If, however, all the parameters of the perturbation – the excessive density, the velocities relative to the unperturbed medium, and the curvature of space – were small, it could not become large by the time its size coincided with the ever growing horizon.

Thus, at early stages of the expansion, when the light pressure dominated, the condensations with sizes greater than the observable horizon – though continuously growing – could not detach and form isolated bodies. The length scale r_{hor}, equal to r_{Jeans}, grew with time and encompassed increasingly large masses, including eventually the condensation that we are considering. What happened to our condensation after that? To understand this, turn to Fig. 29. On this figure the abscissa measures the time since the singularity and the ordinate is the mass of matter inside the perturbed regions. Note that we plot only the mass of ordinary matter and neglect the mass of light, since galaxies form from the ordinary matter. The slanting dotted line in Fig. 29 represents the plot of M_{Jeans} – the mass encompassed at any time by a sphere with radius r_{Jeans} – at the radiation dominated era. (The other lines on this figure will be explained shortly.) The perturbations involving at any given time masses bigger than M_{Jeans} (i.e. those above the dotted line representing M_{Jeans} on the figure) grow, while those with masses smaller than M_{Jeans} (beneath the M_{Jeans} line) do not grow. Now we can trace the evolution of any perturbation with a fixed mass. Draw a horizontal line C corresponding to this mass. At earlier epochs – as can be seen from the figure – our perturbation grows until the time t_1 when the horizontal line C intersects the slanting line representing M_{Jeans}. We already know that at this time the perturbation – despite the previous growth – is still small in amplitude and cannot separate out from the surrounding medium. After a time t_1 the mass of the perturbation is lower than M_{Jeans} and no growth can occur. The perturbation becomes an acoustic wave. It can be shown that the amplitude of acoustic oscillations remains constant – neither increasing nor decreasing – if one neglects the viscous damping. The viscosity is due to the fact that the compressed matter, though highly opaque to light, permits some leakage of photons which gradually squeeze their way out of compressed regions. The viscosity has a different effect on the large-scale and the small-scale perturbations. All small-scale perturbations dissipate. Large-scale perturbations remain intact. More quantitative details of this will be given below.

Dramatic changes occur when the temperature in the expanding Universe drops below 4000 degrees. At this temperature the ionized plasma of the hot cosmic material transforms into a neutral gas (as was described in §9 of Chapter 3). Individual electrons and protons are no longer present – the matter consists of neutral atoms. This event has drastic consequences to the growth of perturbations. The point is that the neutral gas is transparent to light and the latter decouples from the matter. Prior to the decoupling, it was just the pressure of light that determined the elasticity of the medium and caused the Jeans length to be so large. Now light has no effect

Fig. 29. Regions of stability and instability in the hot Universe at different times. For detailed explanations see the text.

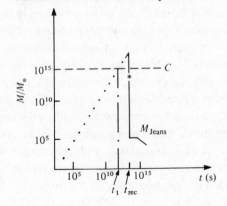

whatsoever on the elasticity of matter – it freely escapes the growing condensations. Only the usual gas pressure is capable of resisting the gravitational forces. But the gas pressure is much less than that of the light, sound propagates much more slowly in neutral matter than light does and, therefore, the Jeans length drops substantially after the decoupling, as is shown in Fig. 29. The gas pressure at the decoupling era was as many times smaller than the light pressure as the number of photons per neutral atom, i.e. by the factor $n_\gamma/n_a \approx 10^9$. Consequently, the sound speed dropped by a factor of $(10^9)^{\frac{1}{2}}$, the Jeans length diminished by the same factor, while M_{Jeans} decreased as the cube of this factor – namely, by $(10^9)^{\frac{3}{2}}$. So, just after the decoupling $M_{\text{Jeans}} \approx 10^5 \ M_\odot$.

Now all the inhomogeneities with masses in excess of $M_{\text{Jeans}} \approx 10^5$ $M_\odot$ can grow due to the gravitational instability. The Jeans length is much less than r_{hor}, all relativistic effects can be ignored and nothing can prevent small condensations from growing large and becoming isolated bodies.

How this process proceeds is the topic of the next section. Here we note only that not all perturbations survived the radiation dominated era. Small-scale perturbations were smoothed out prior to the decoupling due to the viscosity, when they were mere acoustic waves. More detailed calculations show that all the perturbations with masses less than 10^{13}–10^{14} $M_\odot$ had dissipated by the end of the decoupling era. This mass is marked in Fig. 29 with an asterisk.

Summing up the above discussion of the adiabatic perturbations, we wish to emphasize the following. All the region in Fig. 29 to the left of the line $t = t_{\text{rec}}$ is in fact gravitationally stable. No gravitationally bound systems could evolve from initially small perturbations by this time. And this is the answer to the question, posed on page 123, as to why the matter had not fragmented into individual bodies at the very beginning of the expansion. For this reason the dotted line representing M_{Jeans} on the left-hand side of Fig. 29 should be interpreted in a rather Pickwickian sense. It does not actually separate the regions of stability and instability but, rather, marks the boundary between the region where condensations make many oscillations as acoustic waves (to the right of this boundary) and the region where condensations have no time to make even a single oscillation (to the left of this boundary). The true boundary of the instability region is the solid line representing M_{Jeans} on the right-hand side of the figure.

Now we turn to the other types of perturbations.

The simplest to analyse is the behaviour of the entropy perturbations. Before the decoupling these perturbations can neither grow nor dissipate. The inhomogeneities of the ordinary matter sprinkled over the homogeneous photon gas (the relict radiation) remain fixed with respect to the photon gas and expand together with it. Following the decoupling, matter frees itself from the bonds of the radiation and the density inhomogeneities begin to grow under the action of gravitational instability – if their masses exceed $M_{Jeans} \approx 10^5 \ M_\odot$ of course.

Somewhat more complicated is the analysis of vortex perturbations. First of all, consider how the velocities of macroscopic motions in these perturbations change during the course of the expansion of the Universe. The quantity conserved in rotational motion is the angular momentum which can be estimated as Mrv, where M is the mass of a perturbed region, r is its size and v is the vortex velocity. Substituting into this expression $M \approx r^3 \rho$ and recalling that the mass density of the relict radiation varies as r^{-4} (see § 5 of Chapter 3), we conclude that the conservation of the angular momentum implies the constancy of the vortex velocity v. Thus, for the entire period when the matter density is dominated by light – i.e. practically until the moment t_{rec} – the vortex velocities do not change. This conclusion is valid when no viscous dissipation occurs. For the viscous damping of the vortex velocities, however, the same arguments that have been formulated for the adiabatic perturbations apply. Calculations show that all the vortex motions involving masses below $4 \times 10^{11} \ M_\odot$ dissipate by the recombination epoch t_{rec}.

Prior to the decoupling, vortex motions generated no density inhomogeneities. When the decoupling is completed, the velocity of sound drops, as demonstrated above, by a factor of $(10^9)^{\frac{1}{2}}$ and the vortex velocities become supersonic. The ensuing shocks generate density enhancements which subsequently grow due to the gravitational instability as outlined above.

There is one more important feature of the vortex perturbations worth mentioning. Let us trace the behaviour of these perturbations back into the past, towards the singularity. From Fig. 29 one immediately sees that early enough any perturbation exceeded the observable horizon in size. Therefore, one must account for the relativistic gravitational effects due to rotational velocities of the perturbed masses. These relativistic effects turn out to be so strong near the singularity that they drastically change the very character of the cosmological expansion. In other words, for the vortex perturbations to be present at later stages, the cosmological expansion at the very beginning must have been different from what Friedmann's model reveals. This fact alone leads to a conclusion that it is rather unlikely that it was the rotational perturbations that gave rise to galaxies. True, the Soviet astrophysicist G.V. Chibisov has shown that the above unpleasant predicament near the singularity can be avoided if one postulates that the whirls of photons were exactly balanced at that distant epoch by the counter-whirls of gravitons, so that the net vorticity was exactly zero. But this picture does not seem probable, though it is quite possible in principle.

4 **The pancake theory and other theories**

We turn now to the processes which come into play after the recombination. With the light decoupled from the matter, the gravitational instability is capable of developing to its final stage – the formation of isolated celestial bodies. Here we review briefly various theories of this process. We begin with a remark pertaining to all theories. Whichever type of perturbation prevailed at the beginning of the expansion of the Universe, the amplitude of these perturbations by the end of the decoupling era was quite small on scales corresponding to present-day galaxies and larger. As we have already mentioned, the principal evidence for this comes from the observations of the relict radiation. When recording relict microwave radiation, we are actually 'seeing' the decoupling era since which the Universe has been transparent to relict photons. If adiabatic condensations were present at the decoupling era, when the Universe was becoming 'clear', the relict radiation would have had a somewhat higher temperature inside such condensations, and today's terrestrial observers would 'see' them as small enhancements of the intensity of the relict microwave background in the directions of the condensations. The intensity of the relict radiation is also affected by the velocities of cosmic matter at the decoupling epoch – no matter whether these were vortex velocities or the velocities of acoustic oscillations. Such velocities alter the radiation intensity via the Doppler shift. One more factor affecting the intensity of the relict radiation is the gravitational red shift in the gravity fields generated by the matter perturbations. The most accurate observations of the relict microwaves have revealed that their intensity is constant over the sky to an extremely high degree of accuracy. No deviations from the uniform intensity that could be associated with the primordial fluctuations have been found so far, although as high a sensitivity as $\Delta I/I \approx 10^{-4}$ was achieved. From this we conclude that all types of perturbation were very small indeed during the decoupling era. Nevertheless, after decoupling, the gravitational instability would have triggered the growth of these perturbations and been quite capable of amplifying them up to the formation of gravitationally bound objects. This is the process we are going to discuss now. We begin with the adiabatic theory developed largely by the Soviet physicist Ya.B. Zel'dovich and his collaborators.

Recall that this theory postulates small primordial adiabatic perturbations. By the recombination epoch the viscous friction smoothed out all the perturbations with masses below $10^{13} M_{\odot}$. Hence, after decoupling the gravitational forces amplified these condensations with masses $10^{14} M_{\odot}$ and larger. Of highest amplitude, presumably, were the condensations with masses just around $10^{13} M_{\odot}$. All these masses greatly exceeded $M_{\text{Jeans}} \approx 10^{5} M_{\odot}$, so the pressure forces were insignificant to the subsequent growth of these perturbations (see §1 of Chapter 4). As we shall see in a moment, this circumstance leads to a rather peculiar shape for the first objects in the Universe.

To gain a better understanding of the situation, consider a somewhat advanced phase when the gravity forces had already accelerated the particles of the contracting matter in a condensation to some finite velocity. Look what happens next. Earlier, it was believed that clumps would form with more or less spherical shapes. This turned out to be wrong. Actually, oblate two-dimensional structures – which have been called 'pancakes' – originate from this sort of a collapse. How does this happen?

Take some volume element in a contracting condensation. In general, the contraction velocity components are different along different directions (Fig. 30). Along some direction the expansion may even continue, while in other directions the element is already contracting. At any rate, there is always a direction within the chosen element along which the contraction proceeds fastest. Assume for a moment that we have 'switched off' the gravity forces. What occurs then? As we mentioned before, the pressure forces are inefficient and the particles of matter will move by inertia with constant speeds. Eventually, the dominating velocity component will flatten our gas element into a pancake-like structure (see Fig. 30). Now switch the gravity back on. Will it change the overall pattern of collapse? Only slightly. When the initial condensation flattens out, the gravity forces increase but not significantly. In this respect a one-dimensional collapse differs radically from a spherically symmetric contraction for which the gravity forces grow without limit! Thus, although the gravity forces increase the contraction velocities somewhat at the latest stages of the collapse, they inflict no crucial changes on the magnitudes and directions of these velocities and do not alter the overall picture of the 'pancake' formation. In adjacent elements similar processes take place.

Thus, the first objects in the Universe – according to the adiabatic theory – were two-dimensional 'pancakes'. For this reason the theory itself has been called the pancake theory. Fig. 31 shows 'pancakes' obtained in a computer simulation of the above described process.

The masses of the first 'pancakes' born were of the same order of magnitude as the masses of the perturbations with the largest amplitudes at the end of the decoupling era, i.e of the order of $10^{13} M_\odot$

Fig. 30. The contraction of a mass element after the decoupling in the adiabatic theory. (*a*) The initial stage of the contraction; velocities are different in different directions. (*b*) The final stage of the formation of a 'pancake'.

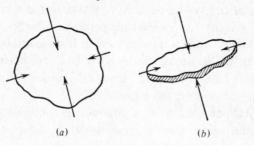

(*a*)   (*b*)

or 10^{14} $M_\odot$ for slightly different values of the parameters). Such masses are typical of an average cluster of galaxies. Thus, the newly born 'pancakes' appear to have been none other than protoclusters which later evolved into the present-day clusters of galaxies. What was this evolution like? For the latest stage of the contraction into a pancake-like structure, when the matter density was high, one can no longer ignore the pressure forces. At this stage a stand-off shock-wave develops and subsequent layers of onfalling material pass through that shock. On passing through the shock front, the gas heats up and becomes turbulent. The further evolution of this gas is determined by the interplay of various cooling and heating mechanisms. Analysis shows that in the central regions of a 'pancake' the gas cools down and breaks into fragments with galactic masses, and afterwards into subfragments of stellar mass within each galaxy. The peripheral tenuous parts of a 'pancake', heated up by the shock, do not have enough time to cool, and remain in a gaseous form – thus becoming the progenitor of the hot intra-cluster gas.

Fig. 31. The formation of 'pancakes' in computer simulations. The 'pancakes' are seen edge-on as lines along which particles group together.

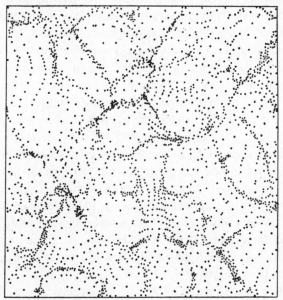

The theory of adiabatic perturbations assumes no initial vortex motions. The gravitational forces by themselves impart no rotation to the perturbations. Where does then the rotation of galaxies, observed in spirals and irregulars, come from? The vortex motions in the pancake theory develop after matter passes through the shock front and becomes turbulent.

Finally, when were the pancakes and galaxies formed? Most likely, it took place at an epoch when the red shift z was somewhere between 10 and 4.

So, these are the general outlines of the process which – according to the adiabatic theory – culminated in the formation of clusters of galaxies, galaxies and stars.

Now we consider briefly how the other theories depict the formation of the structure of the Universe.

According to the entropy theory, which was first discussed by Ya.B. Zel'dovich and his colleagues and later by American astrophysicists R.H. Dicke and P.J.E. Peebles, the clumps of matter initially sprinkled throughout the radiation sea do not dissipate but remain 'intact' up to the decoupling era. No oscillations, no random motions which could cause viscous friction occurred prior to the recombination. Therefore, no dissipative mechanism was in operation and the perturbations remained 'frozen'. After the decoupling all gravitationally unstable condensations with masses $M > M_{Jeans} \approx 10^5$ $M_\odot$ began to grow. Less massive perturbations presumably had larger amplitudes. Hence, the first to become isolated must have been the blobs with $M \approx 10^5 \, M_\odot$. In contrast to the adiabatic theory, of primary interest here are the condensations with masses comparable to the Jeans mass, for which the pressure forces can never be neglected. The pressure tends to level off the anisotropy of the contraction and make the ensuing isolated blobs spherical in shape. Thus, the first objects according to this theory must have been spherical gas clouds with masses typical of globular star clusters. Dicke and Peebles suggested that the first to have formed were the globular clusters which shortly afterwards fragmented into individual stars. It was only later that the globular star clusters gathered into groups and formed galaxies in this way, the latter aggregating subsequently into clusters of galaxies.

Such is the usual version of the entropy theory.

We turn now to the vortex hypothesis. The first work in this field was done by C.F. von Weizsacker and G. Gamov; the theory has been developed to its present level by the Soviet astrophysicists L.M. Ozernoi, A.G. Chernin, G.V. Chibisov and others. According to the most plausible version of the vortex theory, the rotational velocities of the order of $0.1\ c$ and lower – involving masses of the order of 10^{13} $M_\odot$ and larger – survived to the decoupling era.

After the decoupling these velocities generated density inhomogeneities which, in their turn, grew due to the gravitational instability. In this way the first objects in the Universe formed, and in the process of further fragmentation and aggregation, led to the present picture of galaxies and clusters.

Lately, theoretical models have been discussed, which combine the above mentioned theories. One such model has been proposed by the English theoretician M. Rees.

Thus, the reader can see that the problem of the origin of the galaxies is far from being solved yet. To make a judicious choice from among the various theories, one has in the first place to compare their predictions with the observational data.

To this end, it is very important, for example, to perform observations of the relict radiation with greater accuracy. A substantial increase in sensitivity here can be achieved by carrying the receivers out into space, on board cosmic vehicles. The attempts to observe young galaxies in the process of their formation may prove to be even more fruitful. Clearly, for this one has to peer into the deepest regions of space from whence the light and radiowaves, emitted at the era of galaxy formation, are reaching us today. Various means of performing such observations at radio, visible and X-ray frequencies have been suggested. Undoubtedly, new important results will be obtained in this field before long.

5   If the neutrino rest mass is not zero . . .

The whole of the above discussion on the origin of galaxies was based on the assumption that the Universe is practically devoid of particles with a significant total mass, generating strong gravitational fields, and, at the same time, only weakly interacting (via non-gravitational forces) with the ordinary matter and escaping detection in this way (see §§2 and 3 of Chapter 4). Without compelling evidence (either theoretical or experimental) for the presence of such particles, the majority of astrophysicists tacitly assumed that these particles simply did not exist.

The suspicion that our picture of the Universe probably lacks some essential fragments arose many years ago, when the so-called 'problem of the hidden mass' (discussed briefly in §9 of Chapter 1) was formulated. Recall that the essence of this problem is as follows. The galaxies in clusters of galaxies move in such a manner that one is compelled to assume the presence of some invisible mass in the space between galaxies, which through its gravity affects the motion of bodies but does not reveal itself in any other way. Such invisible mass seems to surround giant galaxies as well – which can be inferred from the motions of nearby dwarf galaxies and other objects. This hypothetical invisible mass has been called the hidden (or missing) mass. Observations show that about 20 times as much mass must be hidden within the clusters as is contained in their galaxy members. While the sum of the masses of all galaxies in a typical cluster is about 3×10^{13} $M_\odot$, the total mass of invisible matter has been estimated to amount to 10^{15} $M_\odot$. Nothing else is known as to the nature of this matter.

In spring of 1980 a group of physicists from the Institute of Theoretical and Experimental Physics, Moscow published their results of many years of experiments which suggested a non-zero rest mass for the electron neutrinos. The most likely value of the rest mass of electron neutrinos deduced from these experiments is $m_{0\nu} \approx 5 \times 10^{-32}$ g. This means that electron neutrinos are not bound to move always with the speed of light but, rather, may travel with any speed (below that of light, of course) or be at rest. The importance of this discovery both for physics and, as we shall see below, for astronomy is difficult to overestimate. It should be emphasized however that experiments on the detection of the neutrino rest mass are extremely hard to perform. The results of the ITEP group can in no way by considered as the ultimate proof of a non-zero rest mass and will be checked and rechecked in the future. But if the above mentioned discovery is confirmed, its consequences for astrophysics are so dramatic that theoreticians chose not to wait for confirmation and began to actively explore the possible implications. Beside the ITEP group, some other experimental groups have also published their results on the possibility of a non-zero neutrino rest mass – and not only of electron neutrinos (to which the ITEP experiments pertain) but other kinds of neutrinos as well. We shall not dwell any longer on these experiments here but return to astrophysics.

First of all, note that the possible astrophysical consequences of a non-zero neutrino rest mass were discussed long before the ITEP experiments. Pioneering work in this field has been done by the Soviet physicists S.S. Gershtein and Ya.B. Zel'dovich as early as 1966. Important studies have been performed by the Hungarian physicists G. Marx and A.S. Szalay and others.

But this was, so to speak, a theoretical reconnaissance. Only after the ITEP experiment did various theoretical groups undertake a head-on attack of the problem.

The experiment tells us that neutrinos are 20 thousand times lighter than electrons, and 40 million times lighter than protons. How come the theoreticians claim that these, the lightest, non-interacting particles, play so crucial a role in the Universe? The answer is simple: neutrinos (we mean the relict neutrinos) are quite plentiful in the Universe. Per cubic centimetre they are about a billion times more numerous than the protons (see §5 of Chapter 3). Despite the negligible mass of each individual particle, all together they are a major contributor to the mass in the Universe. From the discussion in §5 of Chapter 3 we know that, if the mass of neutrinos were due only to its kinetic energy and the rest mass were exactly zero, the density of the neutrinos in the Universe would be about the same as the density of the relict electromagnetic radiation, i.e. about $\rho_\nu \approx 10^{-34}$ g cm^{-3}. To estimate their density when the rest mass is not zero but rather $m_{0\nu} = 5 \times 10^{-32}$ g, one has just to multiply this value by the number of neutrinos per cubic centimetre of space. The number density of relict neutrinos is somewhat less than the number density of relict photons (why is explained on page 120). Detailed calculations show that the total number density of electron neutrinos and anti-neutrinos is $N_\nu \approx 150$ cm^{-3}. Now one can easily calculate the neutrino matter density in the Universe to be $\rho_\nu \approx 5 \times 10^{32}$ g $\times 150$ cm$^{-3} \approx 10^{-29}$ g cm^{-3}, which is about 30 times more than the density of the ordinary matter in the Universe! So, the gravity of neutrinos does indeed dominate and governs the kinematics of the cosmological expansion today. By mass, the ordinary matter comprises some 3% admixture to the gas of neutrinos. That is why we can say that the Universe consists mostly of neutrinos, and we actually live in a neutrino Universe (certainly if the value $m_{ov} = 5 \times 10^{-32}$ g is confirmed in future experiments).

The above conclusion has yet another implication.

One of the most important questions in the theory of the evolutionary Universe is whether it will continue to expand forever. The answer to this question (see §7 of Chapter 1) depends on the specific value of the mean matter density in the Universe. If the matter density exceeds the critical value ρ_{crit}, the gravity of this matter will eventually slow down the expansion of the Universe and the galaxies will begin to draw together – the expansion will reverse to a contraction. But if the matter density is below the critical value, the pull of gravity is not strong enough to bring the expansion to a halt, and the Universe will continue to expand forever. The critical value of the density is $\rho_{crit} \approx 10^{-29}$ g cm^{-3}. When the main contribution to the matter density in the Universe was considered as coming from the ordinary matter – for which $\rho_{matter} \approx 3 \times 10^{-31}$ g cm^{-3} – it was believed that $\rho_{matter} < \rho_{crit}$ and the Universe was destined to expand infinitely. Now, there are serious grounds for suspicion that already the density of just the electron neutrinos is of the order of the critical value, $\rho_v \approx 10^{-29}$ g cm$^{-3} \approx \rho_{crit}$. One should not forget that, in addition, there are muon and tau neutrinos. Only very weak constraints on their rest masses have been established so far in direct experiments. But there are theoretical considerations and some indirect experimental evidence implying that if the rest mass of electron neutrinos is non-zero, the rest masses of other kinds of neutrinos must be non-zero as well. In addition, the rest masses of other sorts of neutrinos are likely to be no less than those of the electron neutrinos. If we account for the latter fact, we arrive at a value for the mean matter density in excess of the critical value and conclude that in the remote future, many billions of years hence, the expansion of the Universe will reverse to a contraction. This 'strong' conclusion depends on the 'weakest' of particles, the neutrino, participating in weak interactions only! Here once more we make a qualification, however, since the conclusion of a non-zero neutrino rest mass is still a preliminary one, and all its consequences must be considered as preliminary too.

We turn now to the problem of the origin of structure in the Universe. We already know that the gravitation of neutrinos may be most important in today's Universe. Hence, it is the gravitational action of neutrinos which must first be taken into account when analysing the growth of matter inhomogeneities due to gravitational instability.

The overall picture of perturbation growth appears as follows. At the very first instants after the onset of the expansion there were very small random inhomogeneities in the density distribution of the whole of the matter. Just one second after the beginning of the expansion the matter density was already no longer high enough to interact with neutrinos of various kinds and the latter began their free journey in the Universe. But the energy of the neutrinos was still quite high and they travelled with practically the speed of light. Because they travel freely, the small-scale inhomogeneities in the neutrino distribution are smoothed out. Neutrinos have time enough to escape from not too extensive regions of density enhancements and intermingle with neutrinos from other regions. In this way small-scale neutrino density fluctuations cancel out one another. The more time that has lapsed since the singularity, the more extended neutrino inhomogeneities have had time to be smoothed out. This process of damping continued until the neutrinos, losing their energy during the course of the expansion of the Universe, slowed down to velocities appreciably less than the speed of light. This happened 300 years after the expansion began. Since that epoch neutrinos did not have enough time to escape the large clumps, and the latter, having had initially a small excess density, began to grow under the action of gravity until the entire medium broke into separate contracting neutrino clouds.

What were the masses of these neutrino clouds? They are not hard to evaluate. For the first 300 years the density inhomogeneities were being smoothed out and this process had spread to the scale of the order of 300 light years – the distance covered by neutrinos over this period. On larger scales the clumpiness was preserved and later began to be enhanced due to mutual gravitational attraction, giving rise to isolated, gravitationally bound clouds. Hence, the mass of a typical cloud should be of the same order of magnitude as the mass encompassed by a sphere of radius 300 light years at a time 300 years from the start of the expansion of the Universe. One can readily calculate its value. The neutrino matter density at a time $t = 300$ years can be evaluated from eq. (9) of Chapter 3 which gives $\rho_v \approx 10^{-13}$ g cm^{-3}. The mass of the sphere with radius $r_1 \approx 300$ light years is

$$M_{vcloud} \approx r_1^3 \rho_v \approx 10^{15} \ M_\odot.^\dagger$$

What shape do the condensing clouds take? Since the neutrinos are comparatively slow moving by the time a large density contrast develops, the pressure is of no importance in this process. And as we saw in §4, individual clouds collapse in this case into 'pancakes'. Intersections of nearby pancakes result in a sort of invisible neutrino honeycomb-like structure.

Thus, a cellular structure of invisible neutrino clouds must have formed in space by the present era.

And what about the ordinary matter?

At the earliest stages of the expansion it was, just as the neutrinos, almost uniformly distributed in space. By mass, it constituted only a small fraction as compared to the neutrinos. At a comparatively late stage of the expansion, when the ordinary matter had cooled down significantly, it transformed into a neutral gas. Its pressure dropped sharply. This occurred $\sim 10^6$ years after the singularity. By the end of the decoupling era the cold neutral gas began to condense in the gravitational field of neutrino clouds that had already formed, becoming concentrated in the middle of the neutrino 'pancakes'. Later the clusters of galaxies, galaxies and stars evolved from this condensed gas. Since the ordinary matter accounts for only about 1/30 of the total neutrino mass of the Universe, an invisible neutrino 'pancake' of 10^{15} $M_\odot$ gives rise to a cluster of galaxies with a visible mass 30 times smaller, i.e. some $3 \times 10^{13} \ M_\odot$. This is probably the solution to the problem of the hidden mass – it is hidden in the form of invisible relict neutrinos.

Thus, we see that many features of the large-scale structure of the Universe find a natural explanation in the 'neutrino hypothesis' of galaxy formation described above. We emphasize once more that it depends on the actual value of the neutrino rest mass (or that of some other, as yet unknown particle) as to whether this hypothesis bears any resemblance to reality. So it would be rather premature to draw any ultimate conclusions.

† It is interesting to note that if one performs all the calculations algebraically, in letters, not substituting specific numbers, one arrives at the following formula for this mass, $M_{vcloud} = (c^2 \hbar^3)/(G^3 m_{0v}^2)$, expressed purely in terms of fundamental constants ($\hbar$, Planck's constant, c, the speed of light, G, the gravitational constant) and the rest mass of neutrinos.

5

Frontiers

1 The cosmological singularity

The title of this chapter coincides with the title of the last part of the book by the well-known American physicists C.W. Misner, K.S. Thorne and J.A. Wheeler *Gravitation* (San Francisco, Freeman, 1973). This coincidence is not so much due to the common problems discussed as due to the fact that both here and there questions are raised which lie at the very frontiers of our knowledge, the questions that are just being attacked by contemporary science. One such problem – the problem of the origin of the structure in the Universe – was discussed in the previous chapter. Here we turn to even more intriguing problems – the problem of the singularity, the generalization of gravity theory and others.

We begin with the problem of the initial singularity.

Friedmann's theory inexorably leads to the conclusion that the Universe began from a singular state – the state with an infinitely high density of matter. This conclusion cannot be avoided if one stays within the framework of Friedmann's theory. For example, the following picture is absolutely impossible: first, in some remote past the Universe was contracting; later the contraction, having culminated in a high (but finite) matter density, reversed to the expansion which we observe at present. Such a process would be possible for an isolated body. A massive star, for example, loses stability towards the end of its evolution, begins to contract and develops an enormously high density in its interior. The gigantic pressure of the contracted matter, enhanced by the energy released in nuclear reactions triggered by the contraction, slows down the implosion of the star and reverses it to an explosion. Stellar matter is ejected into space by pressure forces and we observe a supernova outburst. True, one cannot exclude that, in reality, some central part of the star remains bound, but in principle the ultimate expansion of all the stellar material is also possible.

Why is some similar phenomenon not possible for the entire Universe? A tremendous pressure develops in the Universe at high densities just the same, but the pressure by itself does not create any force of expansion – a pressure difference is needed. Only a pressure difference or, as physicists say, a pressure gradient can generate a force. Indeed, imagine a membrane placed in a homogeneous gas without any pressure differences; the gas presses upon the membrane with equal forces on either side and the resultant force is zero. But if the pressure is different on each side of the membrane, the gas presses harder on one of its sides and creates a non-vanishing resultant force acting on the membrane. In the case of a star, there is a high pressure inside the star and a cosmic vacuum outside it, and it is this pressure difference that creates the force of expansion. In a homogeneous Universe there are no pressure differences – the matter is uniform everywhere. There is nothing beyond the Universe (no such notion even exists). Thus, the pressure cannot give rise to a force of expansion in the Friedmannian Universe.[†] This is why the homogeneous isotropically expanding Universe could not have contracted in the remote past, reached a state of very high (but finite) density and then started to expand. In Friedmannian models the expansion starts from a true singularity.

As we argued earlier, Friedmannian cosmology fits the description of the real world from an epoch beginning, most probably, at a fraction of a second up to the present. But maybe the earliest stage of the expansion (prior to a fraction of a second) did not obey the laws of Friedmann's model, and the expansion of the Universe was then anisotropic and the matter distribution non-uniform. If this were the case, perhaps at that non-Friedmannian stage, when pressure gradients and asymmetrical motions of matter were possible, the change from a contraction to an expansion was also possible?

The analogy with the mechanical problem of the expansion of a sphere in Newton's gravitational theory seems to favour such a suggestion. Indeed, if one considers in the framework of Newton's theory the radial outflow of gravitating particles, emerging simultaneously from the centre, one immediately concludes that the expansion began from a singularity. But if one postulates that small

[†] Moreover, Einstein's field equations tell us that the pressure generates an additional gravitational force, i.e. creates a force which does not aid the expansion but rather decelerates it. This circumstance however is of little significance to us here.

random velocities are superimposed on the radial motion of these particles, they miss one another when traced back to the centre, their density is always finite and no singularity arises. Perhaps some analogous situation is possible in the relativistic cosmological problem.

In late 1960s it was shown by R. Penrose, R.P. Geroch and S.W. Hawking that such a possibility can also be excluded. The essence of their work is as follows.

Consider a mass so highly compressed that its gravitational field is very strong and Einstein's equations must be used instead of Newton's law. For this – as we saw in § 12 of Chapter 1 – the mass must be squeezed into a size less than its gravitational radius $r_{\text{grav}} = 2GM/c^2$. Then, following from Einstein's equations, no force in nature can resist the gravitational force, and the matter succumbs to a contraction without limit – a contraction that can never stop and reverse into an expansion; the singularity is unavoidable.

It was shown in § 12 of Chapter 1 that one can always isolate sufficiently large regions in the Universe whose sizes are less than their gravitational radii. Hence, the reversal of the contraction to an expansion in the whole of the Universe would not have been possible in the past, the expansion must have begun from a singularity. The theorems that prove the inevitability of the singularity reveal nothing, however, as to its nature. This is a shortcoming of the theorems. In fact, the theorems do not even claim that the spacetime curvature was infinite at the singularity, or that all the matter was in the state of infinite density. They claim only that the history of at least some particles or photons (assuming that they are not being created or destroyed by interactions) cannot be traced back infinitely in time; that it eventually encounters some singular point in the past. This is certainly very scanty information and the physicists want to know much more.

A prominent achievement in this direction came with the construction of the most general solution to Einstein's field equations near the cosmological singularity by the Soviet physicists V.A. Belinskii, E.M. Lifshitz and I.M. Khalatnikov in 1972. They demonstrated that in the most general case the expansion – if it obeyed Einstein's equations – must have begun as some sort of oscillation with strong anisotropy along different directions.

Does, however, a solution of the most general type pertain to the real situation that occurred in the Universe? As we shall see below, quantum processes must have affected the expansion pattern close to the singularity.

At superhigh densities

$$\rho > \rho_{Pl} = c^5/\hbar G^2 = 10^{93} \text{ g cm}^{-3},$$

the quantum effects become important on the scale of the entire Universe. No theory for such a state is available so far and nobody knows what happened at this density and earlier. It is conceivable that the quantum effects are capable of reversing the contraction to an expansion. If so, a contraction to a density of about 10^{93} g cm^{-3} might have preceded the expansion of our Universe. Whether this was the case is a question that has only just been posed and is still to be solved by modern science.

In §2 of Chapter 2 we discussed some aspects of the contemporary approach to quantum processes in cosmology.

Another possible solution to the problem of the Universe's past is as follows: the matter actually never was in a rarefied state prior to the singularity, and above the density 10^{93} g cm^{-3} the very concepts of space, time, state, etc. lose their meaning; space and time become of a discrete quantum character; the notions of 'earlier' and 'later', and of 'time interval' change drastically. If this is the case, the questions: what preceded the singularity, what happened at, say, minus one second? are incorrectly stated and meaningless. So far, however, these are mere speculations.

Nevertheless, whatever the answers to these new and most interesting questions, there is no doubt that the Universe obeys objective laws amenable to rational scientific study.

2   **The creation of particles in strong variable gravitational fields**

In this section we consider quantum processes which inevitably become important in strong variable gravitational fields near the singularity. These are the processes of the creation of elementary particles. In §2 of Chapter 3 we already evaluated the conditions under which the quantum effects in gravity become particularly strong. They include the earliest stages of the cosmo-logical expansion prior to the moment $t_{Pl} \approx 10^{-43}$ s. (New important phenomena might have occurred at later times and lower densities as well (we have already mentioned the possibility of phase transitions on page 95). We focus our attention here on the epochs when quantum effects in cosmology must have been important in any case.) The method of dimensions enabled us to establish these conditions without much difficulty, but it is not so easy to find out what really happened under such conditions. Important contributions to this end have been made by J.A. Wheeler, R.P. Feynman, B.S. DeWitt, L.D. Fadeev, L. Parker, Ya.B. Zel'dovich, V.L. Ginzburg, M.A. Markov, D.A. Kirzhnitz, A.A. Lubushin, A.A. Starobinskii, A.D. Linde and others. We shall not dwell in detail here on the various aspects of the physical phenomena but, rather, try to explain in a concise and simplified manner how particles can be created from a vacuum under the action of gravitational fields. It should be emphasized that all the discussion will be based on theoretical results only, but even they are not comprehensive enough to resolve the problem. We are still very far from any experimental tests here.

A vacuum in modern physics appears as a 'sea' of so-called virtual particles and anti-particles of various kinds. Unless external fields are present, these virtual particles cannot emerge as real ones. But strong enough or variable fields (such as electromagnetic or gravitational fields) can cause the conversion of virtual particles into real ones.

Theoreticians became interested in processes of this kind a long time ago. Consider the creation of particles by a variable field. It is this sort of process that is important for gravity. Quantum phenomena are renowned as unusual and difficult phenomena to argue from a 'common-sense' point of view. Therefore, we recall one simple example from mechanics before proceeding with the discussion of particle creation in variable gravitational fields. It will help to understand what follows.

Consider a pendulum on a string suspended over a pulley. By pulling up the string or letting it down, one can change the length of the pendulum. If we push the pendulum it begins to swing back and forth. The period of oscillation is determined by the length l of the string:

$$T = 2\pi(l/g)^{\frac{1}{2}}, \tag{1}$$

where g is the free-fall acceleration. Now pull the string up slowly. The length of the pendulum decreases and so does the period of its oscillation. But the swing (the amplitude of the oscillation) increases. Return the string just as slowly to its initial position. The period and the amplitude of the oscillation regain their initial values. Variations in amplitude of this kind are called adiabatic. If one neglects the damping of the oscillations due to friction, the energy of the oscillatory motion in the end remains the same as it was before we started to vary the length of the pendulum. But one can change the length of the pendulum in such a way that, on returning to the initial value of the length, one obtains a different value of the amplitude. For this one has to tug at the string with a double frequency of pendulum oscillations. We do this when we are trying to set ourselves swinging on a swing. This phenomenon is called parametric resonance.

In a similar fashion one can set up electromagnetic waves in resonators. If an electromagnetic wave is excited in a cavity with reflecting walls and a piston, one can change its amplitude by moving the piston to and fro with double the frequency of the wave. By adjusting the phase shift between the piston motions and the electromagnetic oscillations, one can achieve either amplification or attenuation of the electromagnetic wave. But if many experiments are performed with randomly chosen phases, the wave will be amplified on average. Hence, the non-adiabatic character of the process leads to the 'pumping' of energy into the oscillations.

If the resonator contains waves of various frequencies, for any piston motion there will always be a wave with a period of oscillation close to the time scale of the piston velocity change. (Note that in a closed resonator the wavelength values are limited by the size of the resonator. Clearly, the pumping in of energy can occur only if the frequency of the piston motion exceeds the minimum frequency, corresponding to the maximum possible wavelength.) The amplitude of this wave will grow. In terms of quantum physics an increase in amplitude is equivalent to an incease in the number of photons in the electromagnetic wave. So, in non-adiabatic processes new photons – the particles of the electromagnetic field – are born.

Having devoted some time to these simple physical analogies, we return to the vacuum which may be regarded as a 'sea' of various kinds of virtual particles. To simplify matters, we shall discuss only one sort of particle – the virtual photons – but one should bear in mind that the arguments apply to other particles as well. It turns out that non-adiabatic processes, responsible for the amplification of already available oscillations in classical physics, are able to 'amplify' the virtual oscillations too, i.e. to convert the virtual particles into real ones. Thus, temporal variations in the gravitational fields should cause the creation of photons with frequencies of the order of the inverse time scale of such variations. Usually, these effects are completely negligible because the gravitational fields involved are extremely weak. In cosmology, however, near the starting point of the expansion, such effects must have been quite strong due to the immense field strengths and their fantastic rates of change. It is quite possible that quantum processes near the singularity determined the appearance of today's Universe. A systematic study of these effects was initiated by L. Parker in the late 1960s and continued by Soviet scientists. As a more detailed analysis shows, the intensity of particle creation in the isotropic Friedmannian cosmological models strongly differs from that of anisotropic models. In Friedmannian models the particles with zero rest mass (such as photons and, possibly, neutrinos) are not created at all, while heavy particles appear in quite negligible amounts. (The Soviet astrophysicist L.P. Grishchuk has demonstrated that gravitons – in contrast to photons and neutrinos – are created by isotropic expansion too, if the gas pressure deviates from that of the ultrarelativistic particles. The impact of this discovery on cosmology has not been explored yet, but, without any doubt, the role played by this process is by no means small.) A completely different situation occurs in anisotropic expansion. In this case particles are created at an enormously high rate at time $t \approx 10^{-43}$ s. The gravity of newly born particles is so strong that it immediately makes the expansion isotropic!

Perhaps this is just the answer to the question of why the Universe conforms to Friedmann's solution and expands isotropically. But the fact is that the creation of particles, though destroying the anisotropy of the expansion, does not smooth out the inhomogeneity (the inhomogeneity of spatial curvature, for example) of the Universe – if there was any. So, no complete answer to the question of why the Universe is so nicely described by Friedmann's solution yet exists. There is no answer either to the question of why the Universe is hot.

The origin of small perturbations evolving subsequently into galaxies is unknown too. The clue to all these problems is presumably hidden in the quantum processes that took place near the singularity.

3   Matter and anti-matter in the Universe

Throughout the above discussion, we have always assumed that the whole of the observable Universe – the stars, galaxies, etc. – consists practically of matter only, not of anti-matter.

After anti-particles were discovered and the physical theory of the charge symmetry of elementary particles had been formulated, the idea of equal numbers of particles and anti-particles in the Universe, i.e. the idea of a charge symmetry of the Universe as a whole became very attractive. What specifically does this idea speak of? Usually it assumes that some galaxies, or some clusters of galaxies consist of matter, while others consist of anti-matter.

Can such a structure develop? Is it realistic? One of the models of a charge-symmetric Universe was constructed by H. Alfvén and O. Klein. They have developed the ingenious model of the separation of particles from anti-particles by means of magnetic and gravitation fields. But it should be emphasized that the initial state in the Alfvén model is a tenuous plasma out of balance with the radiation. Alfvén rejects the hot cosmological model with a singularity, encountering in this way a difficulty with the explanation of the relict radiation.

A peculiar line of reasoning in the theory of charge symmetry of the Universe is being pursued by R.L. Omnes. It is conceivable – although it has not been proved – that at the earliest stages of the expansion, under very high temperatures, the plasma with large equilibrium amounts of baryons and anti-baryons spontaneously broke down into two distinct phases – one with an excess of baryons, and the other with an excess of anti-baryons. Initially, the regions occupied by the two different phases were small in size but later, according to Omnes, they began to grow.

Omnes invokes a hydrodynamic separation mechanism – the coalescence of regions with a matter excess and, respectively, the coalescence of regions with an anti-matter excess. The physical reason for such a mechanism to operate is the energy release by annihilation at the surface of contact. This phenomenon is usually illustrated with the analogy of the behaviour of a drop of water on a hot stove, i.e. with the motion of the drop during the process of its evaporation – a phenomenon first analysed by J.G. Leidenfrost (1715–94). If the surface of contact separating two phases is curved, a pressure difference develops. The pressure is stronger on the concave side of the distorted surface. Hence, the surface of contact will deform so as to decrease its radius of curvature. In other words, the energy released by annihilation at the contact surface acts as some sort of surface tension tending to decrease the surface area of the contact between matter and anti-matter.

Omnes believes that this mechanism is capable of enlarging the isolated regions of matter and anti-matter to sizes of about 10^{22} cm (with the overall expansion accounted for) by the decoupling era. Such a length scale corresponds to a mass of about 10^{11} $M_\odot$ – the mass of a typical big galaxy.

The process of coalescence is, however, very difficult to calculate and the conclusions drawn do not appear to be particularly reliable. Much more serious are the difficulties encountered by the theory of the charge-symmetric Universe when confronted with the observational data.

According to Omnes' formulae the coalescence process is accompanied by the release of 20 times as much annihilation energy as the relict radiation contains at that epoch. But calculations show that even 1/200 of that energy would be quite enough to noticeably distort the well-studied long-wavelength part of the relict radiation spectrum. Nothing of the kind has been observed.

Thus, the theory of Omnes predicts an energy release hundreds of times in excess of the upper limit compatible with the observations.

In any charge-symmetric theory annihilation must go on in the regions where matter and anti-matter mix. The process of annihilation necessarily includes the characteristic chains of reaction such as $p + \bar{p} \rightarrow \pi^0 +$ other particles, $\pi^0 \rightarrow 2\gamma$, where the energy of the gamma quanta falls in the range 50–200 MeV.

Special surveys in search of such gamma rays gave no results. But the calculations of the gamma-ray background in the various versions of the charge-symmetric theory are not simple. They always require some additional assumptions. A rather serious argument against charge-symmetric theories is as follows. In any such model the regions occupied by matter and anti-matter must be distinctly separated; when approaching either side of the boundary between two such regions the density of both the matter and anti-matter should fall to zero. This means that there is always a contrast in density of the order of unity. The condensation of objects separated from one another began just after the decoupling. But in this case the density of matter in the objects must be of the order of the mean density at the decoupling era, i.e. $\rho \approx 10^{-20}$ g cm^{-3}. The observational data, however, tell us that the density of galaxies is $\sim 10^{-24}$ g cm^{-3}, while the density of clusters of galaxies is about 10^{-27} g cm^{-3}.

Summing up the above discussion of Omnes', Alfvén's and other theories, we conclude that, no matter how ingenious and beautiful, these theoretical models meet such severe difficulties and discrepancies between themselves and the observations that they apparently bear no resemblance to the real Universe. The Universe appears to consist mainly of particles and not of anti-particles. Immediately a question arises: why just of particles? This is one of the unanswered questions. A number of hypotheses have been proposed to explain this fact. All of them invoke quantum processes at the outset of the cosmological expansion. But no satisfactory answer to the question of what was the cause of charge asymmetry in our Universe has yet been found.

4 Can the Universe be oscillating?

We saw in Chapters 2 and 3 that, if the matter density of the Universe exceeds the critical value ρ_{crit}, the Universe is closed; its evolution starts from a singular state, it expands to a maximum size and then begins to contract, ending in another singular state. Whether the real value of the matter density ρ is above or below ρ_{crit} is not known for certain. But let us assume that $\rho > \rho_{crit}$. Is it possible then that the cycles of expansion and contraction follow one another in an everlasting sequence? Can the endless history of the Universe be thought of as being steady-state on average, assuming that its evolution is an infinite sequence of oscillations – the singularity is followed by an expansion which slows down and reverses to a contraction, the contraction speeds up and ends in the collapse? The collapse of the Universe as a whole is assumed to reverse to an expansion in the singular state, starting a new cycle which is a reiteration of the previous one. The corresponding variation of the radius of the Universe is shown in Fig. 32*a*.

The suggested reversal of the collapse to an expansion at the singularity (with quantum effects properly accounted for) is still an open question. We assume that such a transition is possible, and the curve in Fig. 32a on reaching the abscissa is drawn as bouncing up from it.

What will be the consequences of such a model? One feature of the model of the hot Universe seems to favour oscillatory behaviour. Whatever the chemical composition of matter entering the collapse, on passing through the 'fiery furnace' of the singularity it resumes its initial composition, i.e. 70% H and 30% He^4 for the isotropic expansion (see §7 of Chapter 3). This secures a store of nuclear fuel for stars in the new cycle.

In many respects alternating cycles seem quite possible. It is the second law of thermodynamics, however, that forbids the oscillatory model. Indeed, the entropy of the Universe can but grow. The entropy is ever increasing in the course of an expansion as well as in the course of a contraction. But the most significant increase in entropy must occur at the final stage of the collapse. This final stage must proceed according to the most general solution for anisotropic contraction already mentioned in §1 of Chapter 5. Under such conditions the viscosity, the acceleration of neutrinos and the spontaneous creation of particles are particularly intense. And all of these processes result in the increase in entropy.

Fig. 32. The variation of distance with time in the oscillating model of the Universe. (a) Oscillations with an invariant entropy. (b) Oscillations with an ever increasing entropy.

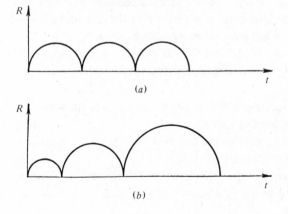

The principal factor in the ensuing argument is the assumption that the entropy does not decrease when the Universe passes through the singularity either. Though we know nothing of physics governing the singularity, we know that the entropy increases in all physical processes studied so far, so it would be only natural to extend this knowledge to the singularity as well. But if the entropy increases from cycle to cycle, every subsequent cycle must differ from the previous one.

Yet it was shown by R.C. Tolman in 1934 that the ever growing entropy of the Universe results in the duration and amplitude of every subsequent cycle becoming larger – the maximum radius of the Universe must increase from cycle to cycle (see Fig. 32*b*).

In every cycle the entropy increases by some finite amount. The duration and the amplitude of the cycles also increase. After an infinite number of cycles the entropy would become infinitely large.

The observational data, however, are not compatible with such a conclusion. Therefore, an everlasting oscillating Universe is not possible under the assumptions made.

Alternative ways of passing through the singularity are also being considered (for example, in the works of J.A. Wheeler, I.L. Rozental and others). These alternative hypotheses admit the change of all the properties of the Universe when passing through the singular state – the values of the fundamental constants of nature, the properties of elementary particles and, sometimes, even the very laws of physics. Under these broad assumptions the entropy, too, can probably decrease.

Of course, the possibility of an infinite number of evolutionary cycles of the Universe cannot be excluded under such a revolutionary approach. What is more, many new attractive possibilities to explain the properties of the Universe arise, but the hypotheses of this kind have not been sufficiently explored yet to be discussed here.

5 Mach's principle in physics and non-Einsteinian theories of gravity

Mach's principle has been discussed in literature since the beginning of this century.

The essence of this principle is as follows: the inertia of a body is determined by its interaction (gravitationally inertial) with other bodies in the Universe. This principle played an important heuristic role in the creation of Einstein's general relativity theory. But after this theory was completed, it became clear that it did not contain Mach's principle! Consider the issue in more detail.

What is the concrete meaning of the statement: the inertia is determined by the interaction with other bodies? A straightforward answer to this question would be that the inertial mass of a body – i.e. the measure of its capacity to resist external forces – is determined by its interaction with other bodies. If no other bodies were present, our test body would have no inertial mass. In empty space no force of inertia (neither Coriolis nor centrifugal forces) would appear in a rotating reference frame. On the other hand, as Einstein wrote, 'the inertia of a body must increase when ponderable masses are piled up in its neighbourhood.'[†]

Neither of the above two statements is true in the theory of relativity. Firstly, empty space obeys the laws of the special theory of relativity in which bodies are endowed with inertia and Coriolis and centrifugal forces act in a rotating reference frame. Secondly, one and the same force – the force of a compressed spring for a example – always imparts one and the same acceleration to a given body regardless of whether heavy masses are or are not present nearby. Therefore, the inertia of a body does not increase 'as weighty masses accumulate nearby'. The opposite statement by Einstein was due to the erroneous interpretation of the formula derived (this error was pointed out by C. Brans and R.H. Dicke). From this point of view any evidence in favour of relativity theory is evidence against Mach's principle.

Things are different, however, with some of the other physical ideas sometimes associated with Mach's principle. For example, Einstein wrote that, in accord with Mach's ideas,

(1) if one accelerates a heavy shell of matter *S*, then a mass enclosed by that shell experiences an accelerative force;

(2) if one rotates the shell relative to the fixed stars about an axis going through its centre, a Coriolis force arises in the interior of the shell; that is, the plane of a Foucault pendulum is dragged around (with a practically unmeasurably small angular velocity).

(taken from W. Misner, K.S. Thorne & J.A. Wheeler, *Gravitation*, Freeman, San Francisco, 1973).

[†] From A. Einstein, *The Meaning of Relativity*, Princeton University Press, 1922.

Both these effects are present in the general theory of relativity. They do not originate however from any variation in the inertial properties of the test body but, rather, reflect the changes undergone by the local inertial (i.e. freely moving) reference frames under the influence of moving gravitating masses. In other words, the mass of a test body is always the same while the inertial reference frame may be different depending on the presence and the state of motion of surrounding bodies. In principle, the effects pointed out by Einstein are of the same nature as the transformation of the inertial reference frame near a gravitating body at rest (near, say, a non-rotating planet). The inertial reference frame falls freely in the gravitational field of such a mass, while it coincides with the inertial frame at infinity when the gravitating mass is taken away. Clearly, such transformations of inertial frames have nothing to do with the inertial properties of a body representing its capacity to resist accelerating (non-gravitational) forces.

So, the general theory of relativity (GTR) does not contain Mach's principle. Therefore, we cannot agree with those who consider GTR to be a theory that confirms Mach's principle. So long as the equations of GTR are believed to be true the controversy is purely scholastic: the outcome of any real or gedanken experiment does not depend on the words with which theoreticians 'explain' the results of their calculations.

Another yet more radical point of view on Mach's principle is sometimes declared. Its advocates considering GTR to be incorrect or, at least, incomplete just because this theory makes no *direct* allowance for the effect of distant masses on the inertia and gravity (rather than through the geometry of spacetime). These authors consider the very fact that GTR admits solutions for empty space as being one of its shortcomings which is to be overcome in future theory.

Any dispute on future theory is a dubious one since no theory yet exists. Nevertheless, not to attack non-existing theories but to defend GTR, we wish to outline the facts that are unfavourable to the critics of GTR.

GTR neither contradicts experiments nor meets any logical difficulties. Hence, there is no compelling objective reason to replace GTR by a 'more Machian' theory.

In our real Universe remote stars and the relict radiation actually define a preferred reference frame, being everywhere at rest (on average) with respect to the matter. Not only the rotation with respect to the relict radiation and remote stars, but the progressive motion as well can be detected and measured – although this motion has no bearing whatsoever on the local laws of nature! In this rest frame we see an isotropic pattern of galaxy recession and measure the same temperature, 2.9 K, for the microwave background in all directions on the sky sphere. An observer in a state of uniform progressive motion with respect to the above defined rest frame would see a temperature higher than 2.9 K for the microwave background in the direction of his motion (and a lower one in the opposite direction) and would register an anisotropic distribution of red shifts for remote galaxies.

Thus, the reference frame associated with the relict radiation and with the overall large-scale mass distribution is truly physically distinguished and is inertial at every point too. Perhaps this fact may be somehow interpreted as confirmation of Mach's principle? We do not think so. A straightforward application of Mach's principle in this context infers the following. Since a preferred reference frame exists, then even the motion by inertia (and not necessarily with acceleration or rotation) must result in the local laws of physics being different in a moving frame from those in the preferred frame. But this is not what occurs in reality – the laws of physics are known to be invariant with respect to Lorentz transformations, i.e. when passing to a moving reference frame. So, Mach's principle, had it turned out to be true, would lead us back to Newton, or even back to Aristotle; one would be able to define a state of absolute rest in the Universe. If experiments prove the Lorentz-invariance of the laws of nature (which they do!), it can be regarded as direct evidence that these laws are independent of any influence from distant masses. In this light the explanation of a limited class of phenomena occurring in accelerated reference frames as being due to the influence of distant bodies loses its attraction. But the author realizes, of course, that considerations of beauty and attractiveness may serve as an heuristic guide at best, and actually prove nothing.

The above arguments were formulated by Ya.B. Zel'dovich and the author many years ago (see the Preface on this). No new grounds for any revision of GTR toward more literal realization of Mach's ideas have appeared since then, although the very thought that the forces of inertia, like any others, should have some concrete source seems rather appealing. Time will show which way the theory will develop.

No attempt to create a 'new' theory of gravity and a 'new' cosmology has been successful yet. The two alternative theories which should be mentioned here are the theory of a 'steady-state' Universe by H. Bondi, T. Gold and F. Hoyle and the gravity theory of C. Brans and R.H. Dicke with its ensuing theory of cosmology. The first theory assumes the continuous creation of matter in the Universe – the creation of new galaxies, gas, etc. (and in today's Universe, not at the beginning of the expansion!) – to compensate for the decrease in density due to the expansion. But the astronomical predictions of this theory have been refuted by direct observational tests.

In addition to the gravitational field (the curvature of the spacetime), the theory of Brans and Dicke postulates the existence of a new scalar ϕ-field whose intensity depends on the distribution of ordinary matter over the whole of space. But here also, the latest observations have put stricter and stricter limits on the possible values of ϕ-field, reducing its allowed range of variation to a negligible level. Accumulated theoretical results in combination with the observational data testify with greater certainty against these theories. Recently, some other attempts to produce alternative models have been also undertaken, but, as we have already mentioned, they are still very far from any degree of success.

The general theory of relativity was created more than 60 years ago on the basis of a minimum number of experimental facts ingeniously picked out by Einstein (see §1 of Chapter 2 on this). For many decades the only experimental tests supporting Einstein's theory were the three famous effects: the deflection of light rays in the gravitational field of the Sun, the reddening of light coming out of regions with stronger gravitational fields, and the slow drift of the perihelion of the orbit of Mercury. Now the number of tests and their precision have increased considerably, but even today we cannot say that all the most important conclusions of GTR have been confirmed experimentally. No wonder that many attempts have been made to construct alternative relativistic theories of gravity (i.e. theories valid at high energies and in strong fields), different from Einstein's. As we have already said, none of these attempts was completely successful. Einstein's general theory of relativity is the only perfectly self-consistent relativistic theory of gravity that is devoid of internal contradictions and conceptually very beautiful. In addition to direct experimental tests, the very analysis of its internal structure, its interrelations with other branches of physics, is quite convincing enough for most researchers that the general theory of relativity is a true one.

The reader interested in a more detailed discussion of the problem is referred to a review paper by N.P. Konopleva, *Usp. Fiz. Nauk,* **124,** pp. 537–63 (1977).

As far as we know, the only absolutely unavoidable generalization of Einstein's gravity theory is the generalization that must incorporate quantum effects in strong variable gravitational fields.

Attempts to generalize GTR in one way or another and to derive a new cosmology from the new theory were and are being made, but, as we have seen above, for no substantial reason. That is why we shall not discuss alternative gravity theories in greater detail here.

6   **The possibility of a non-trivial topology
      for the Universe**
We already mentioned in § 3 of Chapter 2 that the topological properties of the Universe, i.e. the properties of the spacetime as a whole, may not be simple.

The problem of the structure of the Universe as a whole is one of the most imposing problems to have ever challenged the best minds of mankind. Since the beginning of this century when the theory of relativity was created, this question acquired a new meaning and new aspects.

From the times of the ancient Greek and Roman civilizations until the birth of general relativity theory it seemed self-evident that three-dimensional space was Euclidean and time flowed equably everywhere. If this were so, the spatial and temporal structure of the Universe would appear to be quite straightforward: space would be boundless and extend in all directions to infinity, time would have no beginning and no end. Only this kind of world structure seemed acceptable to any 'spontaneous materialist'. The arguments substantiating such a view were probably most clearly formulated two thousand years ago by the great ancient Roman philosopher Lucretius. He wrote in his poem 'De Rerum Natura':

The whole Universe then is bounded in no direction of its ways; for then it would be bound to have an extreme edge. Now it is seen that nothing can have an extreme edge, unless there be something beyond to bound it, so that there is seen to be a spot farther than which the nature of our sense can follow it. As it is, since we must admit that there is nothing outside the whole sum, it has not an extreme point, it lacks therefore bound and limit. Nor does it matter in which quarter of it you take your stand; so true is it that, whatever place every man takes up, he leaves the whole boundless just as much on every side.

Book 1, Verse 960, as translated by Cyril Bailey in his edition of *Titi Lucreti Cari De Rerum Natura*, Oxford, Clarendon Press, 1947.

Since then, arguments of this kind as to the limitlessness and infinity of space were consistently reiterated over the centuries that followed. The infinity of time as mere duration – a river without origin and without end – has never been doubted either. A very clear formulation of this was presented by Newton. So, the structure of space and time as a 'scene' upon which the drama of the evolution of the Universe was taking place seemed quite simple and firmly established.

Before GTR was created, the problem of the structure of the Universe was considered as being the problem of the spatial distribution of matter and fields and their evolution against the familiar scene of Newtonian space and time.

The first blow to this naive concept (from the present point of view) came when a non-Euclidean geometry was discovered by N.I. Lobachevski, J. Bolyai, G.F.B. Riemann and K.F. Gauss. In this respect (as we already mentioned in § 1 of Chapter 2) the profound ideas of the authors of non-Euclidean geometry anticipated GTR, and we shall discuss some of these ideas in the framework of GTR. Thus, it was in GTR that the problem of structure of the spacetime itself first arose.

This problem is very difficult to explore both theoretically and experimentally, and it is not easy to account for in a popular book either. Some difficulties may be attributed perhaps to subjective reasons, due to the fact that we are ready to take up the study of very complicated processes but against the familiar easily understood 'scene' of space and time. When one begins to realize that no familiar 'scene' exists – that its properties are more complicated than those of the processes involved and are actually determined by the latter – some effort is required in accepting it. In this book we are forced to restrict ourselves to the simplest statements and examples, demonstrating, nonetheless, how deep and extraordinary the problem is, of which until quite recently nobody was aware and some prefer not to contemplate even now.

Consider the largest distance scale being currently explored in astronomy. As we saw in §2 of Chapter 2, the curvature of the spacetime must be taken into account over this scale which is of the order of 10^{28} cm. It is the possibility of the significant curvature of the spacetime – i.e. the significant departure of its global properties from those of Euclidean space – from which the question of its topology naturally and inevitably arises. We touched on this question in §3 of Chapter 2 in connection with the homogeneous and isotropic cosmological model for the matter density exceeding the critical value, ρ_{crit}. In this case, three-dimensional space turns out to be closed and finite, though boundless.

Thus, there is at least one case when non-Euclidean topology is unavoidable. From this a straightforward generalization ensues: why not explore possible non-Euclidean topologies when they are not in fact necessary? One of the simplest examples of such a space is illustrated in Fig. 19. It shows a two-dimensional flat manifold with a non-trivial topology, closed in one dimension and infinite in the other. Obviously, a similar example can be constructed for three-dimensional flat space too. More complicated models are also possible. Isolate a parallelepiped in flat three-dimensional space and identify ('stick together', as we did with the paper band in §3 of Chapter 2) its opposite sides. We arrive at a closed flat three-dimensional space. Such a space is called a 'three-dimensional torus'.

Thus, one can assume that there exists a closed flat three-dimensional space. It should be emphasized that the above identifications result in a true closed flat space and not in a periodic structure in infinite flat space. The latter means that there are a limited number of galaxies in the Universe and not an infinite sequence of replications of one and the same set of galaxies filling the whole of the Euclidean space. Such a world has no boundaries and no edges. In this case the observations should reveal many ghost images of any single object, which should be seen simultaneously from different directions at different stages of its history, because light has had enough time to go round this closed world many times since the expansion began. In principle, the situation is analogous to the spherical closed model, with a cosmological constant (see § 5 of Chapter 2), constructed by Lemaître. Recall that a special search for identical objects in opposite directions of the sky, aimed at checking this model, gave no positive results.

An example somewhat more complex than the 'three-dimensional torus' is the so-called elliptical space of constant positive curvature. Cut a sphere along its equator and discard the upper half. We then have only the lower hemisphere left. Now identify ('stick together' opposite points of the equator on the remaining hemisphere (see Fig. 33). We obtain a closed 'elliptical' two-dimensional space. It is almost impossible to imagine this space in a pictorial way despite its two dimensions, but mathematics ascertains that such a manifold is possible! The surface area of the 'elliptical' sphere is one half of the entire sphere's surface. An analogous construction is possible for three-dimensional space of constant positive curvature. For this the dimensionality at every step should be increased by one. Of course, the pictorial image becomes even more difficult to develop. The volume of the elliptical space is one half of that of the spherical space.

As we have emphasized several times already, one should not, when discussing curved spaces, try to imagine them as being actually embedded in some space of higher dimensionality. Nothing exists beyond the Universe. We have not yet discussed the fourth coordinate – time. It must not be forgotten that in fact one has to consider

Fig. 33. Half of a spherical surface after the identification of diametrically opposite points along its equator – *a* with *a'*, *b* with *b'*, etc. – becomes an elliptical sphere.

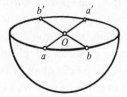

the topology of not only three-dimensional space but that of four-dimensional spacetime.

The above examples demonstrate that a special study of the spacetime topology of the Universe is in order. The number of different topological versions of four-dimensional spacetime is immense and, at first glance, a great variety of options appear in cosmology. For details the reader is referred to the books *Structure of Spacetime* by R. Penrose (Benjamin, New York, 1968), and *The Large-Scale Structure of Spacetime* by S.W. Hawking and G.F.R. Ellis (Cambridge University Press, 1973). In reality, however, even minimal information on the acutal properties of the Universe places severe constraints on the unlimited variety of conceivable topological structures.

Consider some possible topologies of four-dimensional spacetime. We define in this spacetime a time axis. It is common belief that the 'identifications' (similar to those described above for the three-dimensional parallelepiped) along the time axis are forbidden be the laws of physics and must be excluded. This is usually justified by causality considerations. For example, let the time t_2 by greater than the time t_1. If we identify t_2 with t_1, the future and the past can be distinguished only locally, while on the whole, globally, they become mixed up with each another. An interference with physical processes at time t_2 at a point x causes changes at the same point x at time t_1, i.e. in the past.

In fact, the time-like curves may be closed up in much more intricate ways than in the simple example above. For example, some fraction only of all the time-like curves may be closed (in the non-homogeneous case). More information on this can be found in the above mentioned books by Penrose, and by Hawking and Ellis. Here we note only that, in general, the closure of time-like curves is not linked with the causality principle in such a trivial manner. The close of time curves does not necessarily imply a violation of causality, since the events along such a closed line may be all 'self-adjusted' – they all affect one another through the closed cycle and follow one another in a self-consistent way.

In cosmology, however, in view of the spatial homogeneity and irreversibility of the evolution, any time identifications are probably forbidden. Indeed, the specific entropy of the Universe increases continuously with time. And the average abundance of protons decreases continuously due to nuclear reactions in stars. For this reason only, the situation at a later epoch t_2 cannot be identical with that at an earlier epoch t_1.

Having adopted this point of view (which is not the only one possible and is reasonable probably just for a strictly homogeneous Universe), we are compelled to conclude that, as to the time axis, the only question that remains is whether it extends (i) from $-\infty$ to $+\infty$, or (ii) from t_1 to $+\infty$, or (iii) from $-\infty$ to t_1, or (iv) from t_1 to t_2. For strictly homogeneous models the answer to this question can be obtained by integrating the equations of evolution. While current observations reject possibilities (i) and (iii), they do not enable us to choose between (ii) and (iv). Here the 'beginning' and the 'end' of the time axis correspond to the singular states.[†] So, the problem of the topology of the Universe reduces to the problem of the topology of three-dimensional space.

Even so, much freedom still remains in assigning one or another topological type: there are 18 different topologies of flat three-dimensional space and an infinite number (!) of topological versions of isotropic hyperbolic three-dimensional space of constant negative curvature. The analysis for a non-homogeneous Universe (our Universe is non-homogeneous) has not been carried out at all.

Now we turn to the problem of evolution. The most important fact about it is that in homogeneous cosmologies the equations of evolution for local quantities do not depend on the specific topology of the whole of space – the temperature and the density change with time in one and the same manner for any topology.

Hence, the analysis of nucleosynthesis and all other analogous processes in the homogeneous Universe – or, more exactly, in all topological versions of the homogeneous Universe – turns out to be absolutely independent of the topological properties of a particular model. The topology, however, is essential in more global problems.

[†] Recall once more that the question of what 'preceded' the singularity at the outset of the expansion and what will happen 'after' the singularity following the contraction (alternative (iv)) cannot be answered so far in the framework of existing theories. In the immediate vicinity of the singularity the quantum effects of gravity become important and, possibly, the concepts of a continuous metric spacetime fail to work and the notions 'before' and 'after' become meaningless. The exploration of all these issues is a task for the future.

It affects, for example, the overall propagation of signals in the Universe. The appearance of 'ghost' images resulting from multiple round-the-Universe trips of light rays in some topological models is a manifestation of how the topology may affect the propagation of light.

On the basis of this kind of observational prediction one can try to verify in the future the true topology of our Universe. With these brief remarks we conclude our discussion of the topology of the Universe. Bearing various possibilities in mind, it is nonetheless usual to consider the real Universe as having the simplest topology. The word 'usual' will perhaps be substantiated by future research. Currently, this is one of the 'hot' issues in cosmology.

In conclusion, we wish to emphasize that the possibilities discussed above (and others, deeper and more sophisticated, not mentioned here) are not just the fantastic creations of the sophisticated (or, perhaps, delirious) minds of theoreticians. We know that many of the utterly 'crazy' ideas of theoreticians have turned out to be quite real in this century. But it must be remembered that the 'craziness' of science is based not on idle fantasy (still typical, unfortunately, of many of the inventors of half-baked 'hypotheses' who do not trouble themselves with the efforts of acquiring a solid background and do some real work in science) but on sound knowledge and deep insight into the mysteries of nature.

7   Primordial black and white holes

The theory of relativity tells us that not only the topology of the entire Universe may be complex, but that topologically nontrivial structures are also possible on smaller scales. In principle, one can imagine two 'ordinary' spaces connected by a 'throat' – a 'wormhole' making the topology of the total space complex. Such a possibility has been invoked by a number of authors to account for some astronomical phenomena. For example, in 1928 Jeans wrote the following in connection with the theory of the spiral arms of galaxies:

The type of conjecture which presents itself, somewhat insistently, is that the centres of the nebulae (galaxies) are of the nature of singular points, at which matter is poured into our universe from some other and entirely extraneous spatial dimensions, so that, to a denizen of our universe, they appear as points at which matter is being continually created.

(J. Jeans, *Astronomy and Cosmology*, Cambridge University Press, 1928).

Today, the idea of the inflow of matter and energy from some 'other dimensions' through 'holes' connecting two regions of the topologically complex space may seem quite plausible and very appropriate to the explanation of the activity of galactic nuclei and quasars. GTR, however, admits no such possibility. The reason for that is as follows.

Consider such a hypothetical singular region with a 'wormhole'. The equations of GTR written for the outer non-singular region (for a spacetime region far from the centre of a galaxy or a quasar, for example) comply with the laws of conservation: the mass contained in the inner region can change only when some matter or energy (including the energy of gravitational waves) flows out or in across an imaginary surface surrounding this inner region. Any creation of mass and energy, any inflow of these quantities through the 'wormholes' is strictly forbidden in GTR. (When particles are created from a vacuum in a strong gravitational field (see § 2 of Chapter 5), the energy of the gravitational field transforms into the energy of particles and other fields.) Thus, regardless of any fantastic assumptions as to the nature of the singularity inside the surface, one can foretell the exact amount of mass and energy that can emerge from the interior of the surface considered. This amount is determined by the total mass inside the surface which a distant observer can measure by measuring its gravitational field at large distances. Nothing can emerge from the 'hole' that had not previously flowed in through an imaginary surface surrounding the 'hole'.

Thus, Jeans' seemingly attractive idea of a hole leading to 'another world' from which any amount of energy can pour in is absolutely incompatible with GTR.

But the essence of the idea concerning the possibility of a non-trivial topology in small regions proved to be fruitful. It has been embodied in the concept of a 'black hole'. A black hole is a mass so highly compressed that the velocity needed to escape its intense gravitational grip (the velocity that would have to be imparted to a test body to launch it to infinity from the surface of the gravitating mass) exceeds the speed of light. It is quite clear that nothing can come out of a black hole since no object in nature can move with a superluminal velocity The gravitational field attains this enormous strength when a body with a mass M is squeezed into the size of its gravitational radius r_g, given by

$$r_g = 2GM/c^2,$$

which was already mentioned in § 12 of Chapter 1.

In a strong gravitational field, close to the surface of a black hole the geometric properties of space are described by non-Euclidean ('curved') geometry, and time is slowed down as compared to remote regions of space, coming to a 'standstill' at r_g. The region inside r_g may be regarded as a sort of 'hole' in space (and in time). The topology of the interior of the black hole is not known precisely, but it has been established that a spacetime singularity is unavoidable there.

Massive stars end their evolution in gravitational collapse leading to the formation of black holes. What will the formation of such a stellar black hole look like to a distant observer?

An ordinary star, after passing through the violent stages of nuclear flashes and ejection of its outer layers, eventually loses stability and succumbs to a rapid contraction under the action of gravitational forces. But as the surface of a star approaches its gravitational radius, the contraction suddenly slows down due to the slowing down of time. It takes an infinitely long time for the star to reach its gravitational radius – the collapse comes to a standstill. The apparent brightness of the star rapidly decays and its light reddens, since the photons lose energy on coming out of a deep gravitational 'pit', until they are eventually unable to surmount the gravitational barrier at all. A distant observer can never learn what happens to a collapsing star after it contracts beyond its gravitational radius.

Black holes of stellar origin can draw gas from their surroundings, which heats up during the infall and emits X-rays. It is most likely that the first black holes have already been detected as strong X-ray emitters.

The formation of black holes with masses much less than those of stars cannot take place under natural circumstances in the contemporary Universe. In the case of such small masses the gravitational forces are not sufficiently strong to overcome the pressure gradients impeding the contraction, and, for a small mass to be transformed into a black hole, the matter must be compressed to a density much higher than the density of atomic nuclei. Such densities are not encountered in the heavenly bodies of today's Universe.

Recall however that the Universe has been expanding. About 20 billion years ago the density of all matter in the Universe was infinitely higher than its present value. During the first instants after the expansion began the density of matter in the Universe was greater than the nuclear density. At that epoch even rather small density enhancements present in some regions of space were quite sufficient for gravitational forces to reverse the expansion of these regions to a contraction and create microscopic black holes. This process was discussed in § 3 of Chapter 4. In principle, such primordial mini black holes can have arbitrarily small masses. The possibility of the existence of primordial black holes was first pointed out by the Soviet astrophysicists Ya.B. Zel'dovich and I.D. Novikov, and somewhat later by the English physicist S. Hawking.

Did primordial black holes really form in the Universe? What was their fate, how can they be discovered? To answer all these questions, one has to account for the quantum processes which occur in the vicinity of mini black holes. But before going any further in pursuit of this issue, consider one more type of object of the same kind as black holes – the so-called white holes.

The white hole is another theoretical prediction, made by the author of this book and later by Y. Ne'eman. What the physicists have 'invented' here seems even more startling than the properties of black holes. Imagine that the expansion of the Universe did not start simultaneously everywhere but, rather, was delayed for some time in certain regions (cores). In this case singular states would have developed in these regions in the course of the cosmological expansion.

As we have already mentioned, quantum processes become important near singularities. These processes result in the explosive creation of particle–anti-particle pairs in strong variable gravitational fields near the singularity. The cloud of particles created in the vicinity of a delayed core expands and, from the point of view of an external observer, fills the entire region inside the gravitational radius, impeding the expansion of the delayed core. In addition, hot gas falls onto such an object. As a result, the white hole (the matter delayed in its expansion as compared to the rest of the Universe) turns into a black hole.

We shall not dwell any longer here on the description of this interesting phenomenon, but just emphasize once again that even if white holes were present at the beginning of the expansion, and if these delayed cores had not expanded at the earliest stages, they would have quickly turned into black holes.

Now we return to the fate of primordial black holes and to the processes in their vicinity.

The calculations of quantum processes around black holes performed recently by the English theoretician S. Hawking reveal that in the strong gravitational field of a black hole pairs of particles and anti-particles are being continually created from a vacuum at a non-vanishing rate, resulting in the continuous decrease of the black hole's mass and size. Of course, this process is much weaker than those occurring near the singularity. As calculated by Hawking, a black hole of mass M (in grams) creates particles at exactly the same rate as an ideal black body with a temperature of $(10^{26}/M)$ K. On radiating particles, a black hole loses mass and diminishes. Clearly, a black hole evaporates in this manner only if no matter and no radiation fall onto it from outside, because accreting substance, being absorbed by the black hole, will actually increase its mass.

The quantum effects of particle creation are negligible for black holes originating from stars of a few solar masses. A black hole with a mass $3M_\odot$, for instance, has a temperature as low as 10^{-7} K.

Calculations show that, if no external effects interfere, a black hole of stellar mass evaporates for some 10^{66} $(M/M_\odot)$ years (where M is the mass of the black hole). One would have to wait a long time, but it would evaporate all the same!

As the mass of a black hole decreases in the course of its evaporation, its temperature increases, and so does the rate of its evaporation. The last 10^9 g of the black hole mass will radiate for 0.1 s. The energy released, $Mc^2 = 10^9$ g $\times 10^{21}$ cm^2 s$^{-2} \approx 10^{30}$ erg, is equivalent to an explosion of a million megaton H-bombs!

The quantum processes are particularly important for primordial black holes. If at the early stages of the expansion of the Universe, when the matter density was high, black holes with masses below 10^{15} g had formed, they all would have evaporated by the present epoch. For just this reason the process discovered by Hawking is rather important in cosmology. The evaporation of primordial black holes should be accompanied by the emission of high-frequency photons – gamma-rays. Black holes with masses of 10^{15} g should emit gamma-quanta with energies of some 100 MeV.

The detection of such quanta coming from space could, in principle, reveal the existence of primordial black holes. But they have not been discovered so far, and one can only say that the average number of black holes with masses 10^{15} g per cubic parsec of space cannot exceed 10 000. If they were more abundant, the total flux of gamma-quanta with energies of about 100 MeV from these black holes would exceed the presently registered value. The number 10 000 may seem quite large, but one must not forget that the black holes being discussed have negligible masses as compared to, say, typical stars.

Thus, even if primordial black holes really do exist, their total contribution to the mean matter density in the Universe (with each of them weighing 10^{15} g) is negligible compared to that of the ordinary stars. There are other astronomical data imposing severe constraints on the possible amount of cosmic matter in the form of primordial black holes of various masses as well.

Primordial black holes have not been discovered yet. Future observations will show whether these objects really do exist.

8 Concluding remarks

The Universe is evolving; violent processes involving the transformation of matter have occurred in the past, are occurring at present, and will occur in the future. Our present-day view of the Universe differs drastically from the picture of an, 'on average', unchanging Universe, with the same processes occurring always and everywhere in an unchanging space and in uniformly flowing time, which appeared self-evident in ancient times. Everything turned out to be more complicated and more interesting.

Cosmology – the branch of science involving the study of the Universe – is rapidly evolving, bringing more and more new fundamental knowledge about the world that surrounds us.